CONTROL ENGINEERING

CONTROL ENGINEERING

Theory, worked examples and problems

Ruth V. Buckley
Principal Lecturer
Department of Electrical and
Communication Engineering
Leeds Polytechnic

First edition 1976
Reprinted 1978, 1979, 1982, 1983

Published by
THE MACMILLAN PRESS LTD
London and Basingstoke
Companies and representatives
throughout the world

ISBN 0 333 19776 3

Printed in Hong Kong

CONTENTS

PREFACE

The main object in producing this textbook is to provide a series of worked exercises and unworked problems, with answers, in control engineering topics. The general standard aimed at is that of an electrical or mechanical engineering first degree, or the Council of Engineering Institutions examinations in Control System Engineering (Section A of the new syllabus). A small number of the examples are suitable for H.N.C. and H.N.D. courses.

Each chapter is preceded by a brief introduction covering the essential points of theory relevant to the problems that follow and stressing the fundamental principle.

A selection of problems is taken from past examination papers set by the Institution of Electrical Engineers, the Council of Engineering Institutions and Leeds Polytechnic. In this respect, my thanks are due to the Council of Electrical Engineers, the Council of Engineering Institutions and the governing body of the Leeds Polytechnic, for permission to use questions taken from their examination papers, the solutions to which are my own responsibility.

The work is divided into eight main chapters, each dealing with specific aspects of the chosen control section, and an attempt has been made to cover a reasonably wide spread of subject material in the worked examples that follow.

The ninth chapter consists of miscellaneous problems that did not fit directly under the chosen chapter headings but that nevertheless the reader should find helpful in studying control engineering theory.

No attempt has been made to include analog or digital computer examples, since the author feels that these topics are sufficiently specialised to justify a book to themselves, and in fact there are several such good books on the market.

The author wishes to express her gratitude to the publishers for their many helpful suggestions and criticisms and to her colleagues, B. Mann and J.B. Stephenson, for their interest and support during the preparation of the book.

Leeds,
January, 1976

R. V. Buckley

1 TRANSFER FUNCTIONS

When a physical system is being analysed, it is convenient to have a mathematical model in order to examine the features and responses of the scheme. This model can have several forms, among which are the following.

(1) Mathematical presentation, such as differential equations and transfer function relationships.
(2) Graphical presentation in the form of block diagrams and signal flow graphs.
(3) Analog scaled replicas and models.
(4) A program synthesis on a digital computer.

Mathematical representation of the system allows the use not only of differential equations but also of block diagrams, and even though in many cases idealised conditions have to be assumed, much valuable information on the system can be gained from the approximate solutions. This information is valuable because all physical systems are to some extent non-linear and the analysis of such systems would be extremely complex.

Once the physical system is replaced by its *linear* model, the appropriate physical laws may be applied and hence the system equations can be derived.

For electrical systems, the laws would include Ohm's law, Kirchhoff's laws and Lenz's law, while for mechanical systems, there are Newton's laws of motion.

An equation describing a physical system usually involves integrals and differentials, and is called a linear differential equation. Such equations provide a complete description of the system and, for any given stimulus, the output response is obtained by solving these equations. However, this method can be rather cumbersome and difficult for the designer to handle. For these reasons the linear differential equation method is replaced by the transfer function' concept.

The transfer function idea is not a new one; it has been used to describe the Laplace transform relation between the excitation or stimulus of a device and its response. In control work, for a linear system, 'transfer function' is defined as the ratio of the Laplace transform of the output variable from the system to the Laplace transform of the input signal causing that output, all initial conditions being zero. The nature of the signal does not need to be specified.

We shall consider the following two simple examples.

Example 1.1

A circuit consisting of RC elements as shown in fig. 1.1.

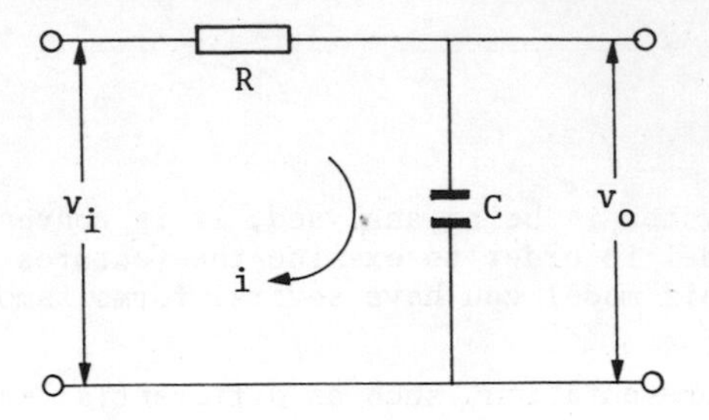

Figure 1.1

Using the differential equation method and applying Kirchhoff's law

$$v_i(t) = Ri + v_o(t) \tag{1.1}$$

$$v_o(t) = \frac{1}{C}\int i \, dt \tag{1.2}$$

Taking the Laplace transform on both sides of these two equations and assuming zero initial conditions

$$V_i(s) = RI(s) + V_O(s) \tag{1.3}$$

$$V_o(s) = \frac{I(s)}{sC} \tag{1.4}$$

eliminating I(s) between eqns 1.3 and 1.4, the transfer function of the network is given by

$$\frac{V_o}{V_i}(s) = \frac{1}{1 + sRC} = \frac{1}{1 + sT} \tag{1.5}$$

where T = RC.

Example 1.2

A simple mass-spring-friction system is shown in fig. 1.2. Consider the force P to be the input and the displacement y of the mass M as the output of the system.

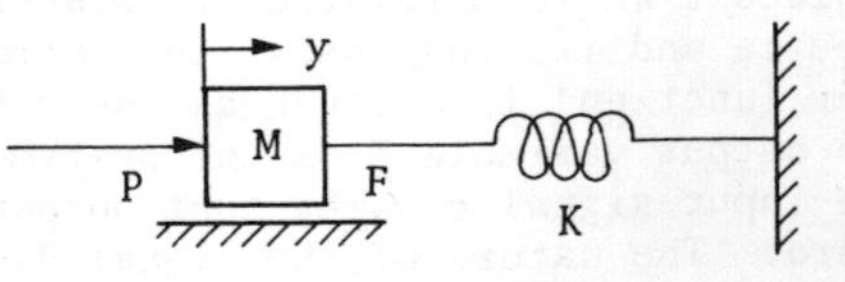

Figure 1.2

The differential equation relating these two variables is

$$P = M \frac{d^2y}{dt^2} + F \frac{dy}{dt} + Ky \tag{1.6}$$

Again take the Laplace transform on both sides of eqn 1.6

$$P(s) = (Ms^2 + Fs + K)\,Y(s) \tag{1.7}$$

The transfer function of the system is given as

$$\frac{Y}{P}(s) = \frac{1}{Ms^2 + Fs + K} \tag{1.8}$$

Example 1.3

Determine the transfer function for the circuit shown in fig. 1.3 assuming no external load.

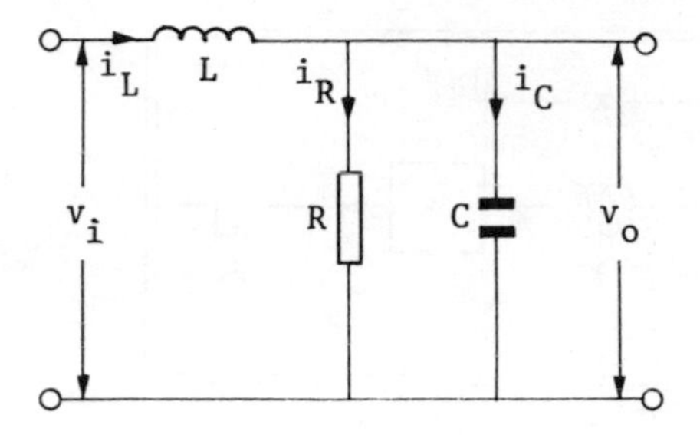

Figure 1.3

Applying Kirchhoff's laws to the above circuit yields

$$v_i(t) = L \frac{di_L}{dt} + v_o(t) \tag{1.9}$$

$$v_o(t) = Ri_R = \frac{1}{C} \int i_C \, dt \tag{1.10}$$

Applying the Laplace transform on both sides of eqns 1.9 and 1.10

$$V_i(s) = LsI_L + V_o(s) \tag{1.11}$$

$$V_o(s) = RI_R = \frac{I_C}{sC} \tag{1.12}$$

also $I_L = I_R + I_C$ (1.13)

therefore

$$V_i(s) - V_o(s) = Ls\left[\frac{V_o}{R}(s) + sCV_o(s)\right]$$

$$\frac{V_o}{V_i}(s) = \frac{1}{LCs^2 + \frac{L}{R}s + 1} \qquad (1.14)$$

Note that L/R and RC have the dimensions of time so that the transfer function may be expressed

$$\frac{V_o}{V_i}(s) = \frac{1}{T_1T_2s^2 + T_1s + 1} \qquad (1.15)$$

Example 1.4

A simple mechanical accelerometer is shown in fig. 1.4. The position x of the mass M with respect to the accelerometer case is proportional to the acceleration of the case. Determine the transfer function between the input acceleration and the output x.

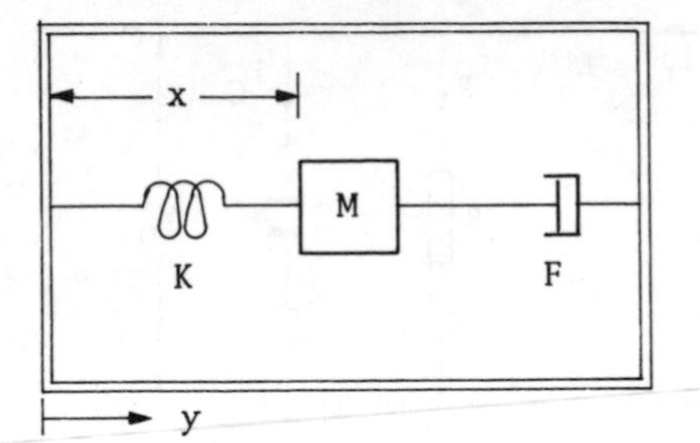

Figure 1.4

In this example, the sum of the forces acting on the mass M is equated to the inertia acceleration. Thus

$$M\frac{d^2(x - y)}{dt^2} + F\frac{dx}{dt} + Kx = 0 \qquad (1.16)$$

therefore

$$M\frac{d^2x}{dt^2} + F\frac{dx}{dt} + Kx = M\frac{d^2y}{dt^2} = Ma \qquad (1.17)$$

where $a = d^2y/dt^2$ = input acceleration. Assuming zero initial conditions, the Laplace transform yields

$$(Ms^2 + Fs + K)\,X(s) = MA(s)$$

$$\frac{X}{A}(s) = \frac{1}{s^2 + \frac{F}{M}s + \frac{K}{M}} \qquad (1.18)$$

It is now thought that the student should tackle problems in the Laplace transform mode without first resorting to the differential equation method, so the subsequent examples are worked out in Laplace form.

Example 1.5

Derive the transfer function of the circuit shown in fig. 1.5.

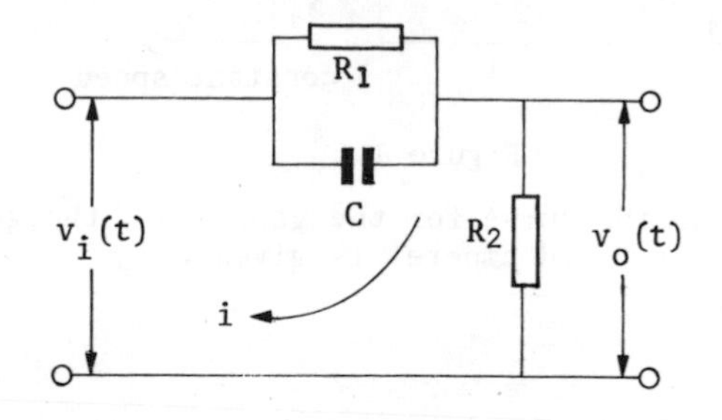

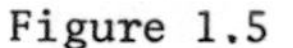

Figure 1.5

$$V_i(s) = \frac{R_1}{1 + sR_1C} I(s) + V_o(s) \tag{1.19}$$

$$V_o(s) = R_2 I(s) \tag{1.20}$$

$$V_i(s) = \frac{R_1}{1 + sR_1C} \frac{V_o}{R_2}(s) + V_o(s)$$

therefore

$$\frac{V_o}{V_i}(s) = \frac{R_2(1 + sCR_1)}{R_1 + R_2 + sCR_1R_2} \tag{1.21}$$

Eqn 1.21 is often rearranged to give the standard form of a phase lead network used in compensation work for control systems.

$$\frac{V_o}{V_i}(s) = \frac{\alpha(1 + sT)}{1 + s\alpha T} \tag{1.22}$$

where $\alpha = R_2/(R_1 + R_2)$ and $T = CR_1$.

Note The simple circuit obtained when $R_1 = 0$ is not used in control systems since it would block d.c. signals and could not therefore be used in the forward path of the system.

Example 1.6

Fig. 1.6 shows a d.c. shunt wound generator, rotating at constant speed, with a signal voltage applied to its field winding. Derive the transfer function.

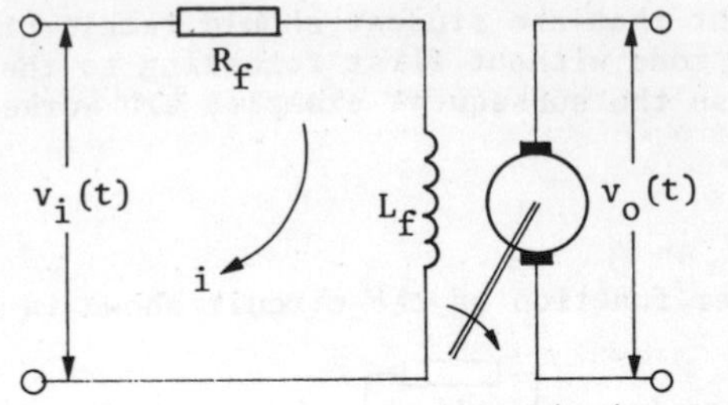

Figure 1.6

From a magnetisation curve for the generator, the generator constant K_g (volts per field ampere) is given by

$$K_g = \frac{V_o(s)}{I_f(s)} \tag{1.23}$$

From the electrical circuit

$$V_i(s) = R_f I_f + sL_f I_f \tag{1.24}$$

therefore

$$\frac{V_o}{V_i}(s) = \frac{K_g}{R_f + sL_f} = \frac{K}{1 + sT_f} \tag{1.25}$$

where $K = K_g/R_f$ and $T_f - L_f/R_f$, the field circuit time constant.

Example 1.7

A four-way hydraulic-type valve controls the flow of oil to a ram driving a load as shown in fig. 1.7. Given that Q_o is the flow through the valve per unit valve opening, V_o is the volume of oil

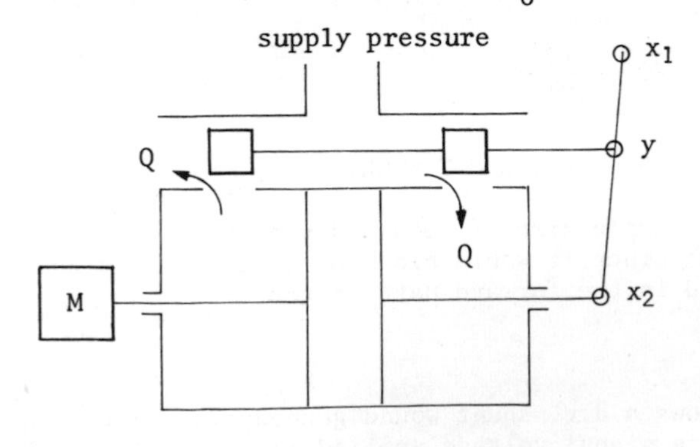

Figure 1.7

in the system, K is the bulk modulus of the oil, M is the mass of the load, A is the ram cross-sectional area, and L is leakage coefficient, derive the transfer function for the system when the actuator position is fed back to the valve by means of a one-to-one linkage.

The flow Q into the main cylinder has three functions: it is used to move the piston Q_v, leak oil past the piston Q_L and create oil compression in the cylinder Q_c.

$$Q = Q_v + Q_L + Q_c \tag{1.26}$$

Now $Q_v = Asx_2$ (1.27)

and $Q_L = L\,\delta p$

where δp is the pressure difference across the main piston and causes the load to accelerate.

$$Ms^2x_2 = A\,\delta p \tag{1.28}$$

Hence

$$Q_L = \frac{LM}{A}\,s^2x_2 \tag{1.29}$$

The definition of bulk modulus is K dV = V dp where dp is the pressure change causing a volumetric strain dV/V. But Q_c = dV/dt and since $V_o/2$ is the volume of one half of the cylinder, then dp/dt = $(2K/V_o)Q_c$. The same rate of pressure change occurs in the other half of the cylinder so that the rate of change of pressure differential is given by

$$\frac{d}{dt}\,\delta p = \frac{4K}{V_o}\,Q_c \tag{1.30}$$

Hence from eqn 1.28

$$Q_c = \frac{MV_o}{4KA}\,s^3x_2 \tag{1.31}$$

Now $Q = Q_o y$ (1.32)

so that eliminating Q between eqns 1.26, 1.29 and 1.31

$$y = \frac{A}{Q_o}\,sx_2 + \frac{LM}{Q_oA}\,s^2x_2 + \frac{MV_o}{4KAQ_o}\,s^3x_2 \tag{1.33}$$

From the one-to-one linkage $y = \frac{1}{2}(x_1 - x_2)$, therefore

$$\frac{x_2}{x_1}(s) = \frac{1}{\frac{MV_o}{2KAQ_o}s^3 + \frac{2LM}{AQ_o}s^2 + \frac{2A}{Q_o}s + 1} \tag{1.34}$$

A more accurate analysis would take into account the fact that the oil flow Q depends on the pressure difference across the spool and hence on δp. However, δp is usually small compared with the supply pressure.

Example 1.8

The following data refer to a d.c. servomotor.

Moment of inertia = 4×10^{-4} kg m^2

Coefficient of viscous friction = 10×10^{-4} N m rad^{-1} s

Torque constant = 2.4 N m per field ampere

Obtain an expression for the transfer function of the motor in terms of the field current and the angular velocity of the output shaft.

Motor torque = acceleration torque + viscous friction torque

$$2.4I_f(s) = 4 \times 10^{-4}\, s\omega_o + 10 \times 10^{-4}\, \omega_o \tag{1.35}$$

therefore

$$\frac{\omega_o}{I_f}(s) = \frac{2.4}{4 \times 10^{-4}\, s + 10 \times 10^{-4}} = \frac{2400}{1 + 0.4\, s} \tag{1.36}$$

Thus 0.4s is the mechanical time constant of the motor.

Example 1.9

A process plant consists of two tanks of capacitance C_1 and C_2. If the flow rate into the top tank is Q_3, find the transfer function relating this flow to the level in the bottom tank. Each tank has a resistance R in its outlet pipe. (Consider the tanks to be non-interacting.)

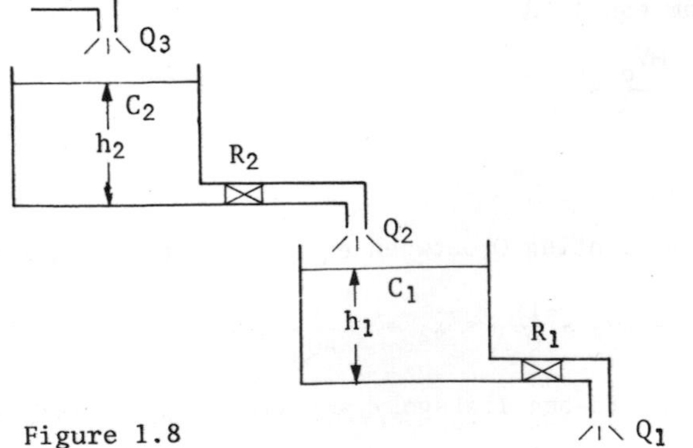

Figure 1.8

Reminder: Q = height/resistance, capacitance is cross-sectional area.

For the upper tank

$$Q_3 - Q_2 = C_2sh_2 \quad (1.37)$$

therefore

$$s(Q_3 - Q_2) = C_2s^2h_2 \quad (1.38)$$

or $$s(Q_3 - Q_2)R_2 = R_2C_2s^2h_2 \quad (1.39)$$

But $$R_2sQ_3 - sh_2 = R_2C_2s^2h_2 \quad (1.40)$$

therefore

$$h_2 = \frac{R_2}{1 + R_2C_2s} Q_3 = \frac{R_2}{1 + sT_2} Q_3 \quad (1.41)$$

where $T_2 = R_2C_2$.

For the second tank

$$C_1sh_1 = Q_2 - Q_1 \quad (1.42)$$

leading to

$$h_1 = \frac{R_1}{1 + sT_1} Q_2 \quad (1.43)$$

where $T_1 = R_1C_1$. But $Q_2 = h_2/R_2$, therefore

$$h_1 = \left[\frac{R_1}{(1 + sT_1)(1 + sT_2)}\right] Q_3 \quad (1.44)$$

Example 1.10

The following data refer to a two-phase induction motor.

Rated fixed phase voltage = $115V_1$

Stalled motor torque at rated voltage = 50×10^{-3} N m

Moment of inertia of rotor = 1.83×10^{-6} kg m^2

Viscous friction of motor = 70×10^{-6} N m rad^{-1} s

No-load speed = 4000 rev/min

Determine the position transfer function and hence the motor time constant.

Assume a linear torque-speed characteristic and let k be the stalled rotor torque at rated voltage. Then

$$k = \frac{50 \times 10^{-3}}{115} = 0.435 \times 10^{-3} \text{ N m per volt} \tag{1.45}$$

Let B be the slope of the linear torque-speed curve, which is a negative number. Then

$$B = -\frac{\text{stalled rotor torque}}{\text{no-load speed}} = -\frac{50 \times 10^{-3} \times 60}{4000 \times 2\pi}$$

$$B = -119 \times 10^{-6} \text{ N m rad}^{-1}\text{ s} \tag{1.46}$$

therefore

$$\text{motor torque } T_m = kV_1 + Bs\theta_m$$

$$T_m = 0.435 \times 10^{-3} V_1 - 119 \times 10^{-6} s\theta_m \tag{1.47}$$

also $T_m = Js^2\theta_m + Fs\theta_m$

$$= 1.83 \times 10^{-6} s^2\theta_m + 70 \times 10^{-6} s\theta_m \tag{1.48}$$

therefore

$$\frac{\theta_m}{V_i}(s) = \frac{2.3}{s(1 + 0.97 \times 10^{-2} s)} \tag{1.49}$$

The motor time constant is 0.0097s.

PROBLEMS

1. A separately excited d.c. generator has the following open-circuit characteristics.

E.M.F. (volts)	3.5	60	117	170	200	233
Field current (amps)	0	2.5	5	7.5	10	15

The field circuit has an inductance of 2.5 H and a resistance of 250Ω. Derive the transfer function relating open-circuit armature voltage to field-circuit terminal voltage for the linear portion of the characteristic.

$$\left[\frac{0.088}{1 + 0.01s}\right]$$

2. Fig. 1.9 shows a small speed-control system and its parameters. The generator is driven at a constant speed and the motor is provided with a constant field current. Further data are as follows.

Generator e.m.f. e_g = 1500 V per field ampere

Motor e.m.f. e_m = 1.0 V rad^{-1} s

Motor torque = 0.32 N m per armature ampere

Inertia of motor and load $J = 0.48 \times 10^{-4}$ kg m^2

Friction is negligible

$r_1 = 1000\Omega$, $L_1 = 10$ H, $r = 100\Omega$

Find an expression relating the instantaneous angular velocity ω_o rad s^{-1} of the load to the input voltage v_i.

$$\left[\frac{\omega_o}{v_i}(s) = \frac{1.5}{(1 + s0.01)(1 + s0.015)}\right]$$

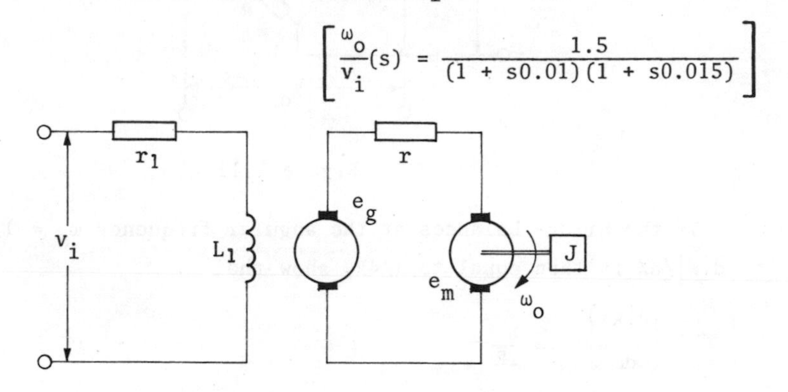

Figure 1.9

3. The circuit shown in fig. 1.10 is used in an amplifier of a control system. Derive an expression for the transfer function of the circuit. If $V_i = 10 \sin 10t$ volts, $R = 50$ kΩ, $R_o = 5$ kΩ and $C = 1\mu$F, calculate the output voltage in magnitude and in phase relative to V_i.

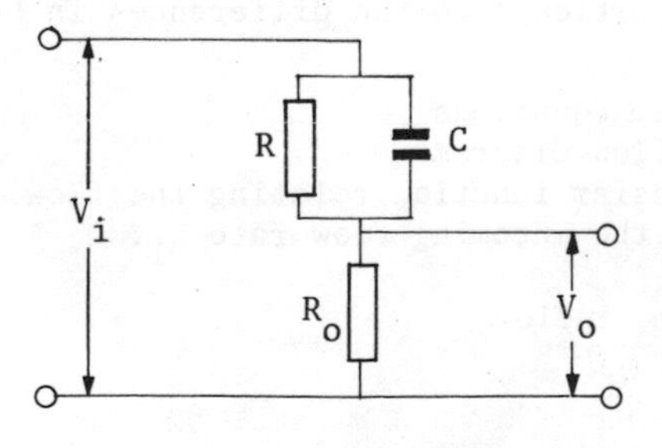

Figure 1.10

(I.E.E. Part 3, 1965)

$$\left[\frac{R_o(1 + sRC)}{R + R_o\ (1 + sRC)};\ 1.015\text{ V } 24° \text{ lead}\right]$$

4. If X is the reactance of the LC combination in the bridge network shown in fig. 1.11, derive expressions (in terms of X and R) for (a) the magnitude and (b) the phase of the voltage transfer function, K, of the network.

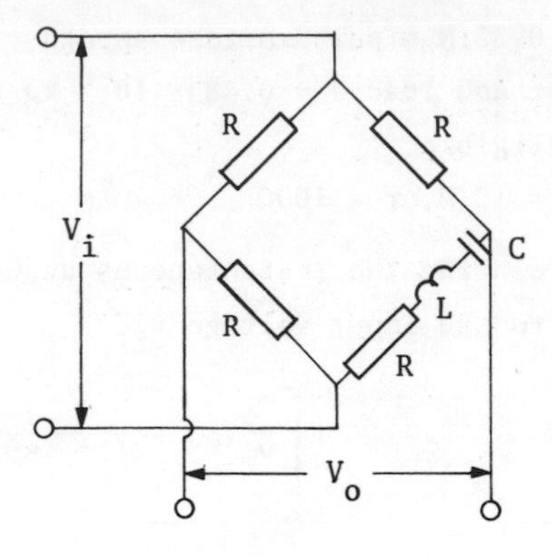

Figure 1.11

If the bridge balances at the angular frequency $\omega_o = 1/\sqrt{(LC)}$ and if $d|K|/dX$ is then equal to $1/4R$, show that

$$\left(\frac{d|K|}{d\omega}\right)_{\omega_o} = \frac{L}{2R}$$

(I.E.E. Part 3, 1964) $\left[|K| = \frac{X}{2(X^2 + 4R^2)^{\frac{1}{2}}},\ \tan^{-1} \pm \frac{2R}{X}\right]$

5. In the hydraulic system shown in fig. 1.12 two cylindrical tanks are connected by a pipe containing a valve. The flow discharges through a valve from the right-hand tank. The depths of liquid in the tanks are H_1 and H_2 respectively. A varying flow rate Q_1 is discharged into the first tank. Assuming that the flow rates through the valves are proportional to the differences in head across the valves

(a) write the system equations
(b) draw a signal flow diagram
(c) derive the transfer function relating the flow rate from the right-hand tank to the incoming flow rate Q_i.

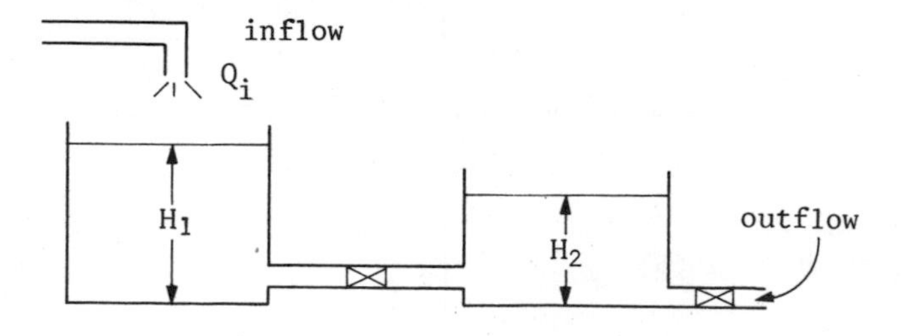

Figure 1.12

(C.E.I. Part 2, Control Systems, 1971)

6. A proportional controller used in process-control plant is shown in fig. 1.13. This device is normally used to position control valves in a process plant system. Derive the transfer function between the pressure P_b to the flapper movement y.

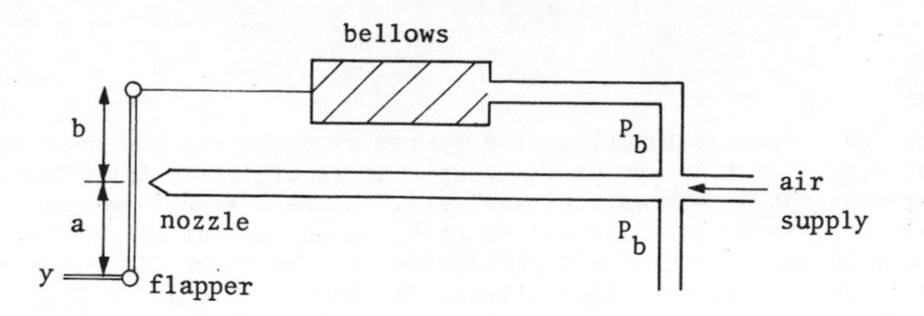

Figure 1.13

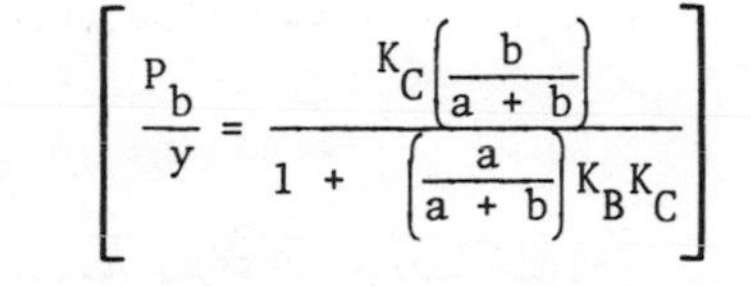

$$\left[\frac{P_b}{y} = \frac{K_C\left(\frac{b}{a + b}\right)}{1 + \left(\frac{a}{a + b}\right)K_B K_C}\right]$$

K_B = constant relating bellows extension to pressure and K_C = gain constant.

7. For the geared system shown in fig. 1.14 find the transfer function relating the angular displacement θ_L to the input torque T_1, where J_1, J_2, J_3 refer to the inertia of the gears and corresponding shafts. N_1, N_2, N_3 and N_4 refer to the number of teeth on each gearwheel.

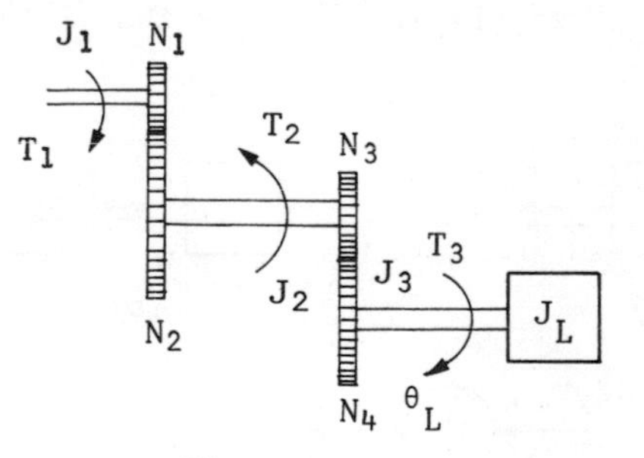

Figure 1.14

$$\left\{T_1 = \frac{N_3 N_1}{N_4 N_2}\left[J_3 + J_L + \left(J_2 + J_1\frac{N_2^2}{N_1^2}\right)\left(\frac{N_4^2}{N_3^2}\right)\right]s^2\theta_L\right\}$$

2 BLOCK DIAGRAMS

Once the transfer function of a system or component has been derived the significant nature of the component is of little importance when carrying out a mathematical analysis. Hence a simple method of analysing a system is to represent it by an equivalent block diagram and then apply several simplifications to the block diagram circuitry. The equivalent block diagram for the RC circuit of fig. 1.1 is shown in fig. 2.1.

$$V_i(s) \longrightarrow \boxed{\frac{1}{1 + sT}} \longrightarrow V_o(s)$$

Figure 2.1

Since most systems have several blocks interconnected by various forward and feedback paths, the algebraic manipulations are more readily carried out if a short-hand notation for the transfer function is introduced. A popular presentation uses G with a suitable subscript.

Fig. 2.2 illustrates the simplest of rules that should be followed to obtain the system transfer function. An individual block representing a transfer function is shown in fig. 2.2a, where $Y = G_1X$. For blocks in cascade, as indicated in fig. 2.2b, $Z = G_1G_2X$; while for a unit that compares signals, such as that illustrated in fig. 2.2c, $Z = X - Y$.

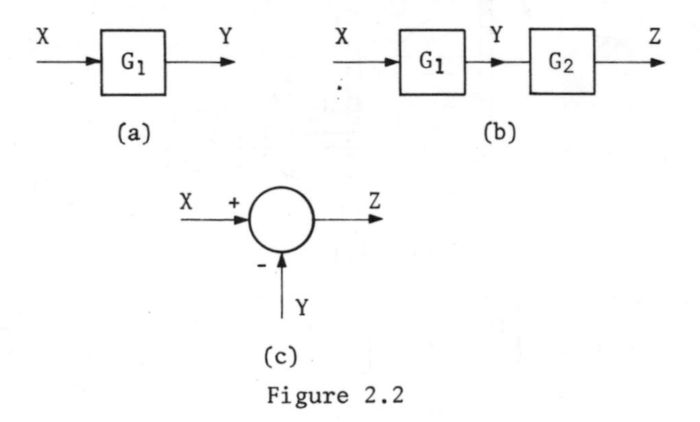

Figure 2.2

Fig. 2.3 shows the standard form of block diagram for a control system with feedback, where $\theta_o = G(s)\theta_e$ and $\theta_e = \theta_i - \theta_o H(s)$.

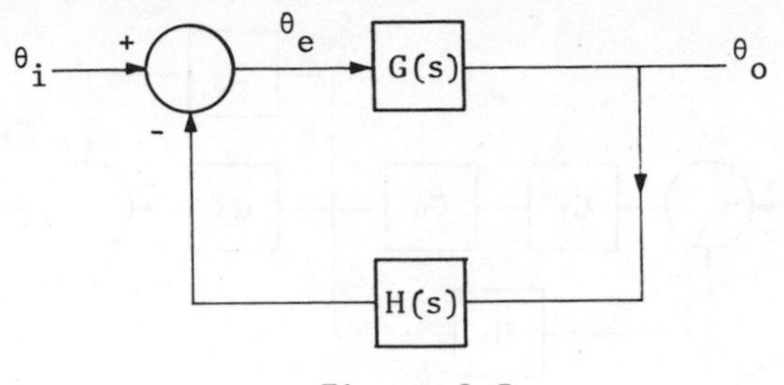

Figure 2.3

Therefore

$$\frac{\theta_o}{\theta_i}(s) = \frac{G(s)}{1 + G(s)H(s)} \qquad (2.1)$$

The characteristic equation of the system is found by taking the denominator of eqn 2.1 and equating it to zero.

$$1 + G(s)H(s) = 0 \qquad (2.2)$$

It can thus be compared with the characteristic equation of differential equation representations. This equation is essential when studying the stability of a system, as will be shown in chapter 5.

Block diagrams of complicated control systems may be simplified using easily derivable transformations; a selection of the more useful ones are given in appendix A.

It is often necessary to evaluate a system's performance when several inputs or disturbances are applied simultaneously at different parts of the system. For *linear* systems only, the principle of superposition may be used (see example 2.2).

The theorem for superposition states that 'the response $y(t)$ of a linear system due to several inputs $x_1(t)$, $x_2(t)$, ..., $x_n(t)$ acting simultaneously, is equal to the sum of the responses of each input acting alone'. Thus if $y_1(t)$ is the response due to the input $x_1(t)$, then

$$y(t) = \sum_{1}^{n} y_n(t) \qquad (2.3)$$

Example 2.1

From the block diagram shown in fig. 2.4a determine the relationship between θ_o and θ_i by successive block diagram reduction.

Take each section and determine the equivalent block.

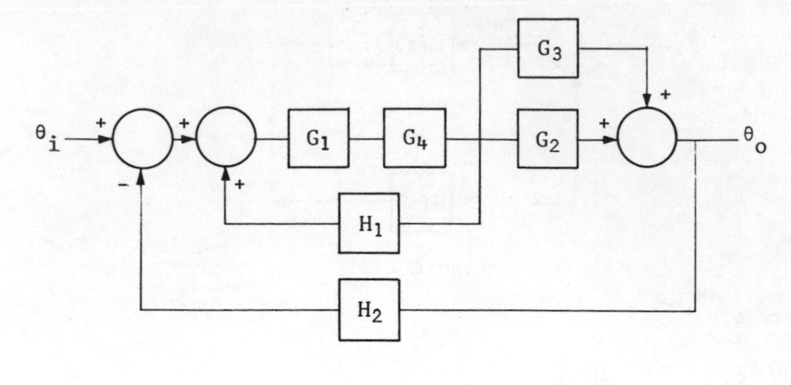
θi
+
+
+
-
G1
G4
G3
G2
+
+
θo
H1
H2

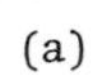
(a)

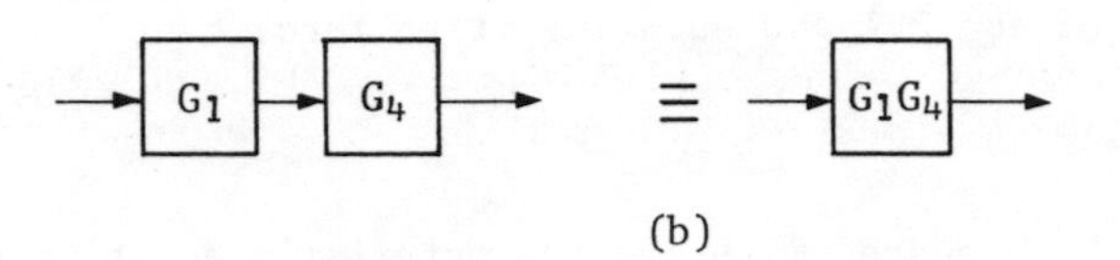
G1
G4
≡
G1G4

(b)

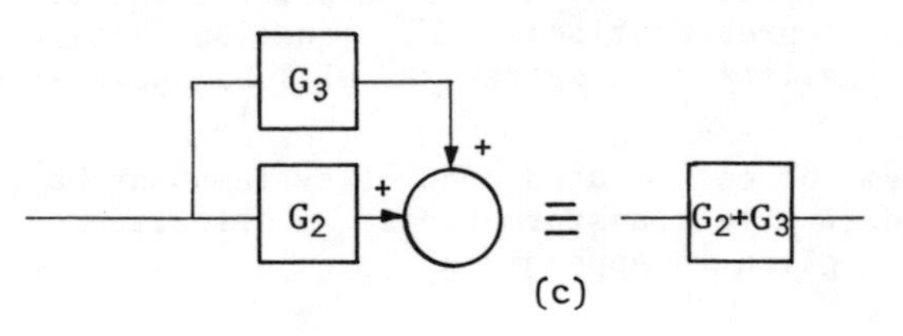
G3
G2
+
+
≡
G2+G3

(c)

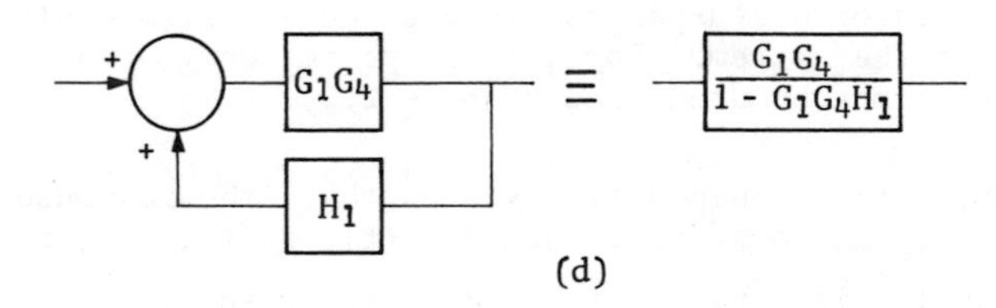
+
+
G1G4
H1
≡
G1G4
1 - G1G4H1

(d)

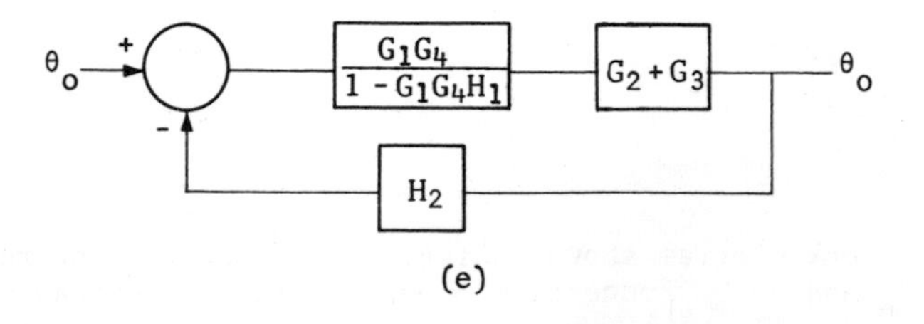
θo
+
-
G1G4
1 - G1G4H1
G2 + G3
θo
H2

(e)

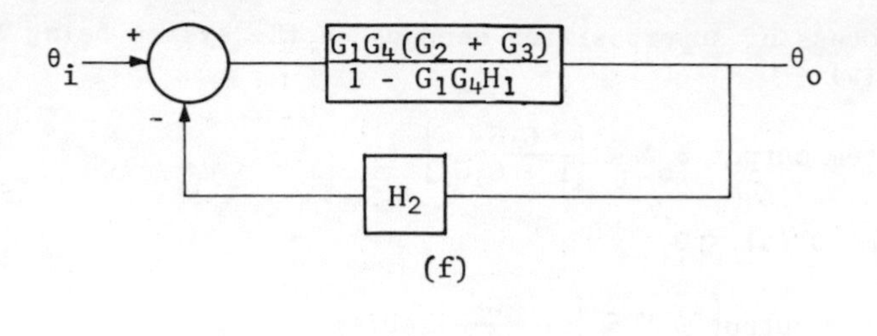

(f)

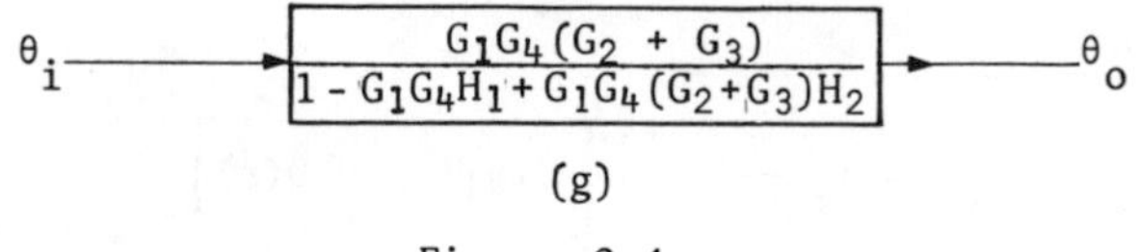

(g)

Figure 2.4

Example 2.2

A closed-loop control system is subjected to a disturbance D(s) as shown in fig. 2.5. Show by the principle of superposition the effect on the output of the system.

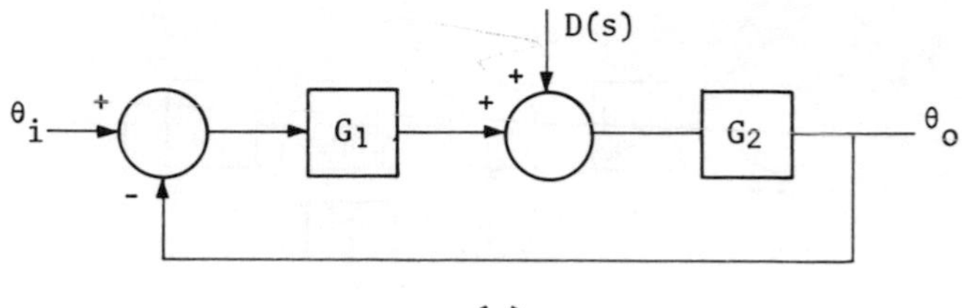

(a)

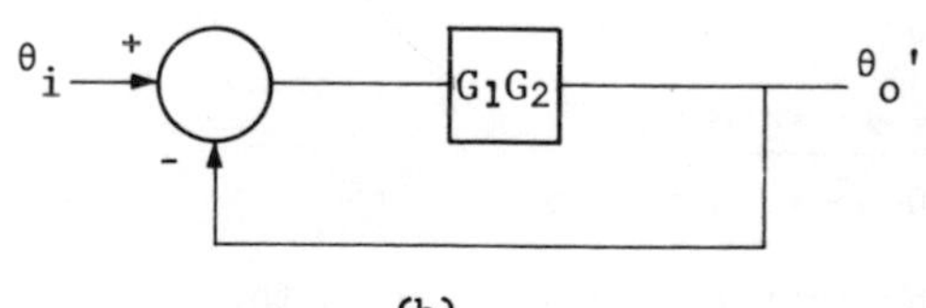

(b)

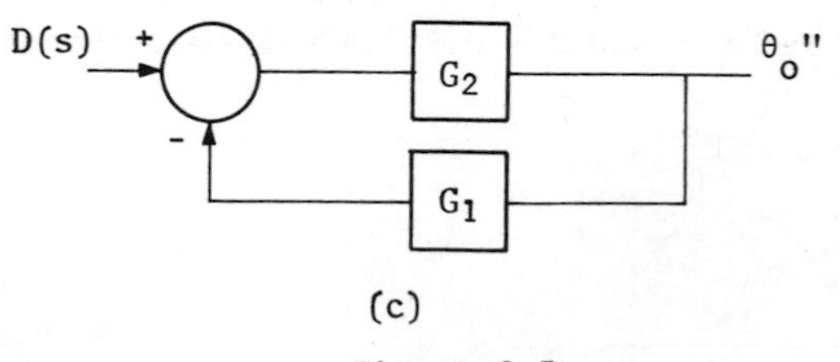

(c)

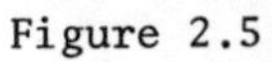
Figure 2.5

The process of superposition depends on the system being linear. Assume $D(s) = 0$.

$$\text{system output } \theta_o' = \left(\frac{G_1G_2}{1 + G_1G_2}\right) \theta_i \tag{2.4}$$

Now assume $\theta_i(s) = 0$.

$$\text{system output } \theta_o'' = \left(\frac{G_2}{1 + G_1G_2}\right) D(s) \tag{2.5}$$

therefore

$$\theta_o = \theta_o' + \theta_o'' = \left(\frac{G_2}{1 + G_1G_2}\right) \left[G_1\theta_1(s) + D(s)\right] \tag{2.6}$$

Example 2.3

In an electrical servo used to control a rotatable mass, the inertia is 100 kg m^2, the motor torque is 1600 N m per radian of misalignment and the damping ratio is 0.5. Develop the block diagram for this system and hence the transfer function relating the position of the output shaft to the input control wheel position.

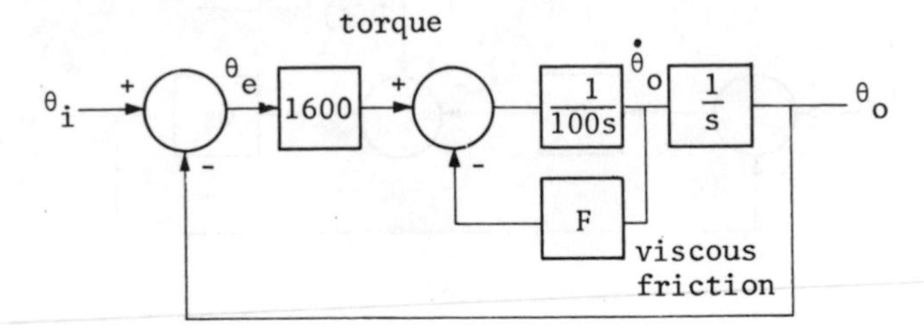

Figure 2.6

From fig. 2.6

$$\frac{\theta_o}{\theta_i}(s) = \frac{1600}{100s^2 + Fs + 1600} \tag{2.7}$$

compare the denominator of eqn 2.7 with the standard form

$$s^2 + 2\xi\omega_n s + \omega_n^2$$

so that $\omega_n = 4$ rad s^{-1} and $2 \times 0.5 \times 4 = F/100$, therefore

$$F = 400 \text{ N m rad}^{-1} \text{ s}$$

therefore

$$\frac{\theta_o}{\theta_i}(s) = \frac{16}{s^2 + 4s + 16} \tag{2.8}$$

Example 2.4

The diagram in fig. 2.7 shows part of a wind tunnel aircraft-pitch-control system. The pitch angle is θ_o and the pilot's input is θ_i,

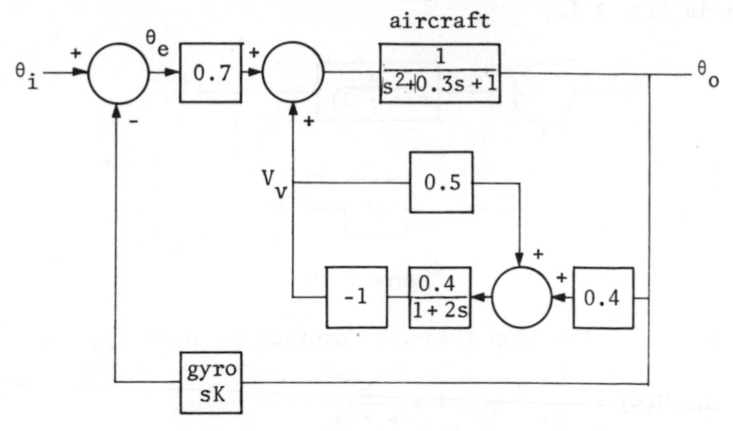

Figure 2.7

the vertical velocity signal is V_v, while θ_e is the elevation angle. By reducing the diagram, determine the closed-loop transfer function.

Fig. 2.8 shows the reduction of the inner feedback path from fig. 2.7, while fig. 2.9 shows a further simplification to the standard form. Eqn 2.9 can then be deduced. Thus

$$\frac{\theta_o}{\theta_i}(s) = \frac{0.7\ (0.6 + s)}{s^3 + (0.9 + 0.7K)s^2 + (1.18 + 0.42K)s + 0.68} \tag{2.9}$$

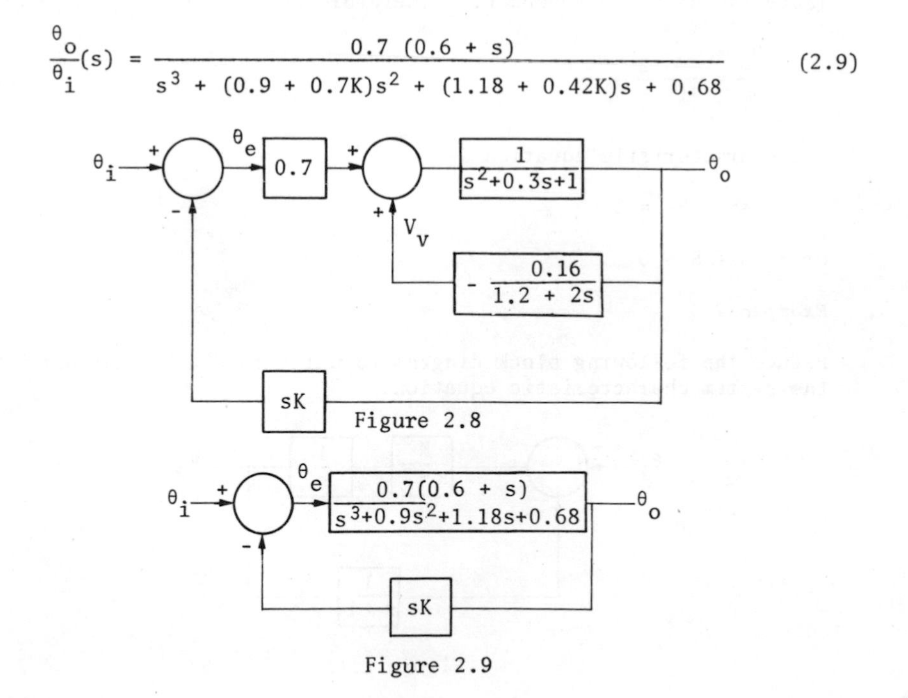

Figure 2.8

Figure 2.9

Example 2.5

Determine (a) the loop transfer function (b) the closed-loop transfer function and (c) the characteristic equation for the system shown in fig. 2.10.

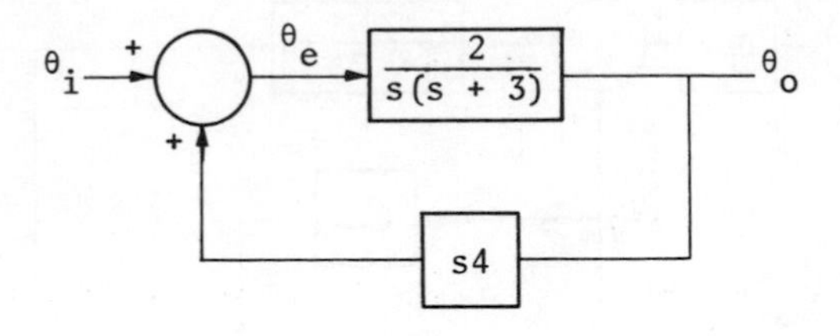

Figure 2.10

From eqn 2.1 the loop transfer function is G(s)H(s), i.e.

$$G(s)H(s) = \frac{2}{s(s+3)}\, s4 = \frac{8}{s+3}$$

while the closed-loop transfer function is

$$\frac{\theta_o}{\theta_i} = \frac{G(s)}{1 - G(s)H(s)} = \frac{2/[s(s+3)]}{1 - \dfrac{8}{s+3}} \tag{2.10}$$

(Note the *positive* feedback.) Therefore

$$\frac{\theta_o}{\theta_i} = \frac{2}{s(s-5)} \tag{2.11}$$

The characteristic equation is

$$s^2 - 5s = 0$$

or $\quad s - 5 = 0$

Example 2.6

Reduce the following block diagram to unity-feedback form and find the system characteristic equation.

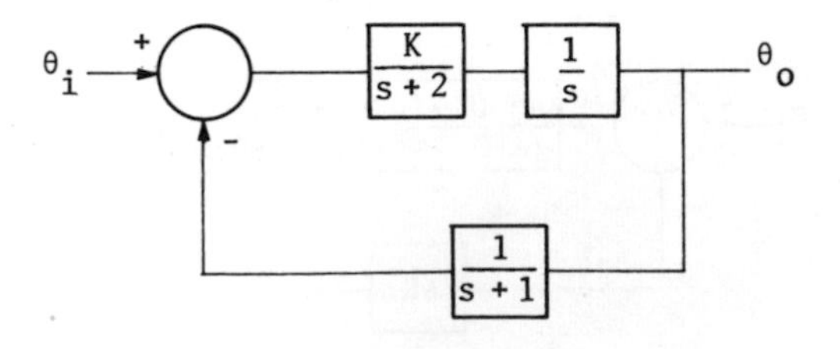

Figure 2.11

Combining the blocks in the forward path, yields fig. 2.12.

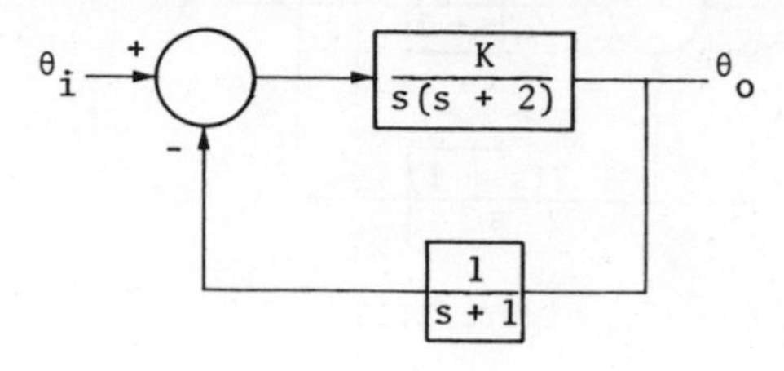

Figure 2.12

Applying transformation 5, i.e. removing a block from a feedback loop, yields fig. 2.13.

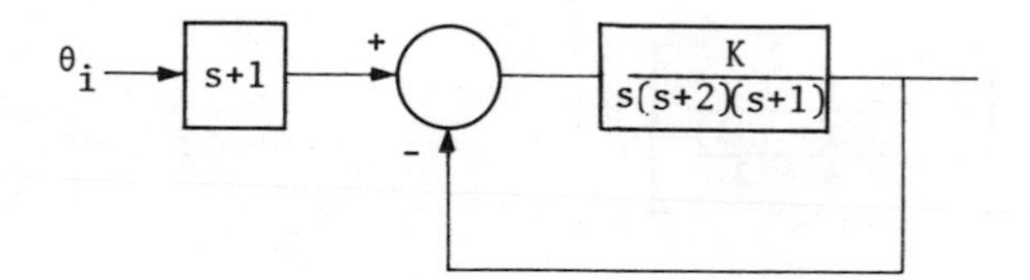

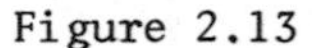
Figure 2.13

The characteristic equation for this system is

$$s(s + 2)(s + 1) + K = 0$$

or $$s^3 + 3s^2 + 2s + K = 0 \qquad (2.12)$$

Example 2.7

Determine the output θ_o for the following system.

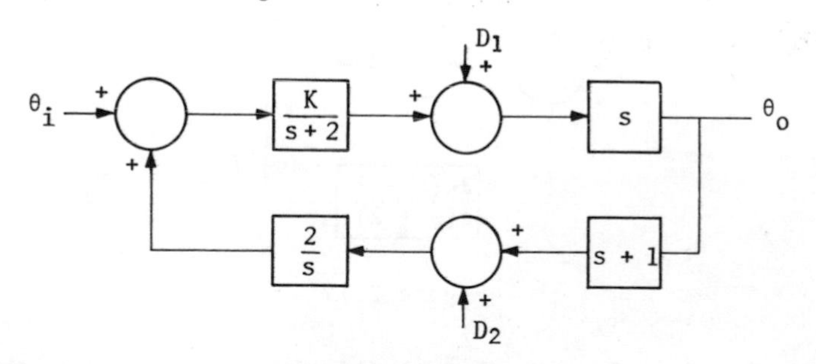

Figure 2.14

Let $D_1 = D_2 = 0$. By combining the blocks in cascade the system becomes as shown in fig. 2.15, where θ_o' is the output due to θ_i acting alone. Therefore

$$\theta_o' = \left[\frac{Ks/(s + 2)}{1 - \dfrac{2K(s + 1)}{s + 2}}\right]\theta_i$$

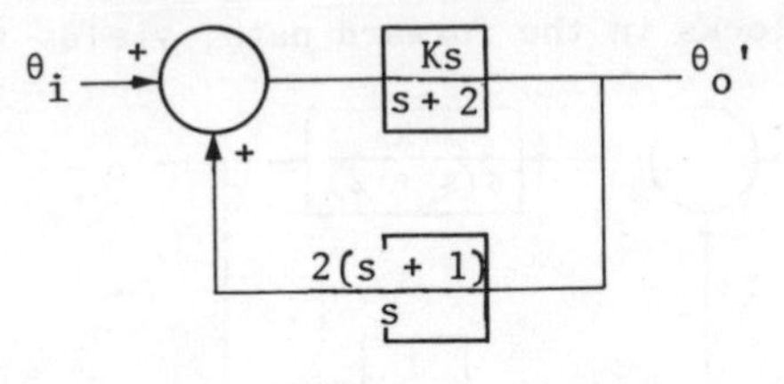

Figure 2.15

Now let $\theta_i = D_2 = 0$. The block diagram becomes as shown in fig. 2.16, where θ_{o_1} is the response due to D_1 acting alone. By rearranging the blocks we have fig. 2.17. Therefore

$$\theta_{o_1} = \left[\frac{s}{1 - \dfrac{2K(s + 1)}{s + 2}}\right] D_1$$

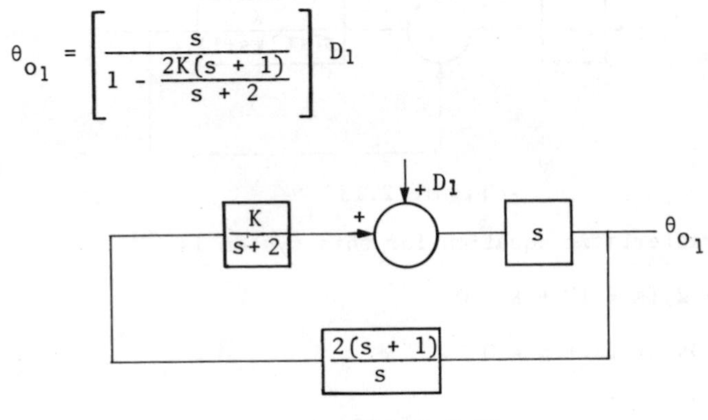

Figure 2.16

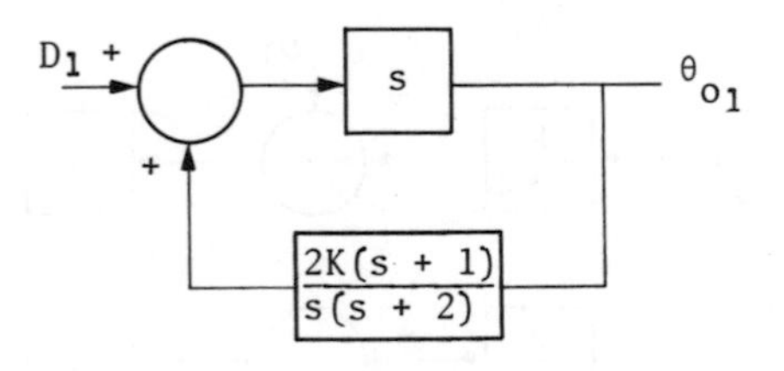

Figure 2.17

Finally let $\theta_i = D_1 = 0$, then the block diagram becomes as shown in fig. 2.18, where θ_{o_2} is the response due to D_2 acting alone. Rearranging the blocks gives fig. 2.19. Hence

$$\theta_{o_2} = \left[\frac{2K/(s + 2)}{1 - \dfrac{2K(s + 1)}{s + 2}}\right] D_2$$

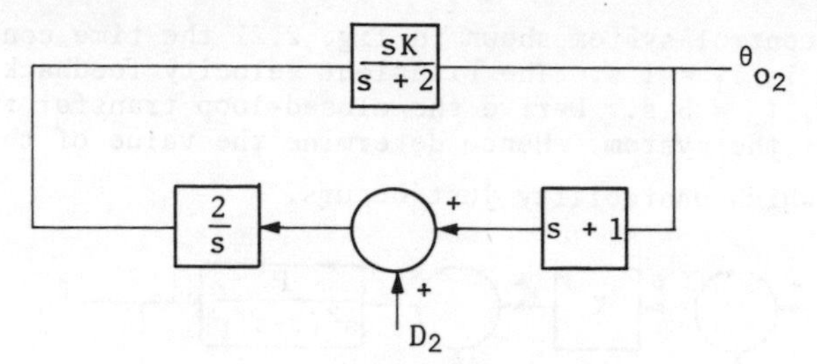

Figure 2.18

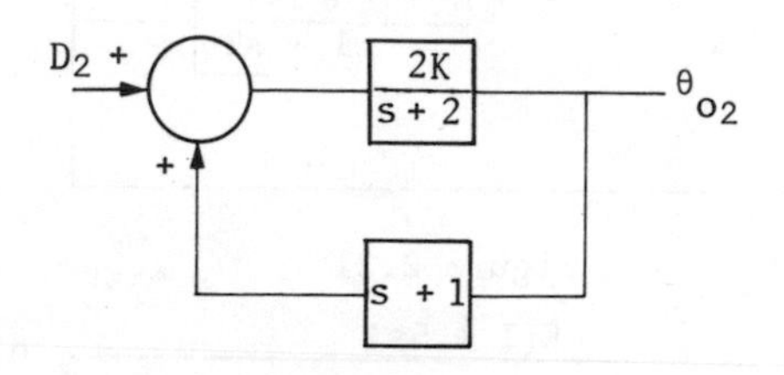

Figure 2.19

By superposition, the output is

$$\theta_o = \theta_o' + \theta_{o_1} + \theta_{o_2} \tag{2.13}$$

$$\theta_o = \left[\frac{Ks/(s + 2)}{1 - \dfrac{2K(s + 1)}{s + 2}}\right]\theta_i + \left[\frac{s}{1 - \dfrac{2K(s + 1)}{s + 2}}\right]D_1$$

$$+ \left[\frac{2K/(s + 2)}{1 - \dfrac{2K(s + 1)}{s + 2}}\right]D_2 \tag{2.14}$$

PROBLEMS

1. Redraw the block diagram of fig. 2.20 to obtain a relationship between θ_i and θ_o.

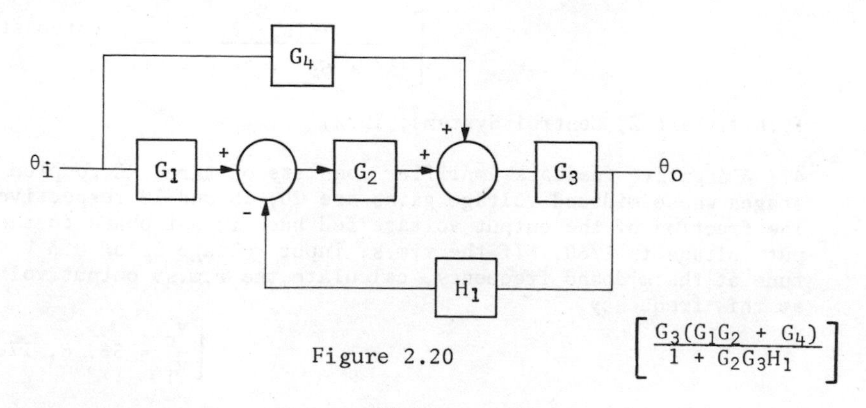

Figure 2.20

$$\left[\frac{G_3(G_1G_2 + G_4)}{1 + G_2G_3H_1}\right]$$

2. In the control system shown in fig. 2.21 the time constant of the power drive is $T_1 = 1$ s. The transient velocity feedback parameters are $\alpha = 0.1$, $T_2 = 5$ s. Derive the closed-loop transfer function $\theta_o/\theta_1(s)$ for the system. Hence determine the value of the scalar gain K for which unstability just occurs.

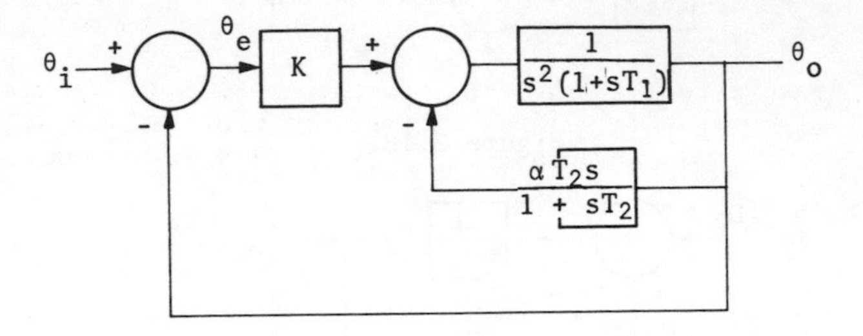

Figure 2.21

$$\left[\frac{K(1 + 5s)}{5s^4 + 6s^3 + s^2 + s\,(0.5 + 5K) + K},\quad 0.048 < K < 0.14\right]$$

3. The diagram in fig. 2.22 is a block diagram of a multiloop control system.

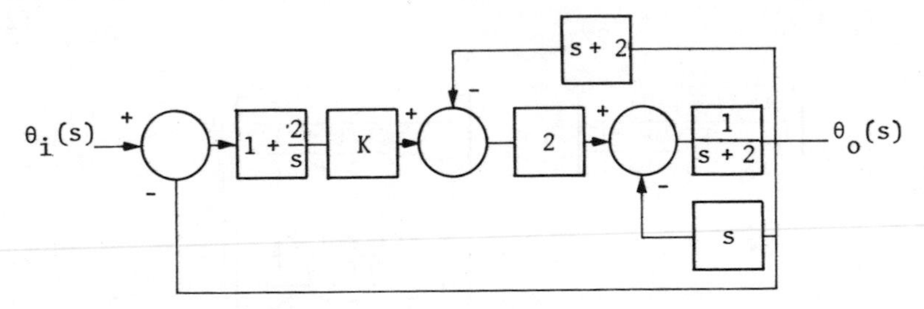

Figure 2.22

(a) Determine the transfer function relating output θ_o and input θ_i.

(b) Sketch the root locus and comment on its significance.

$$\left[\frac{2K(s + 2)}{4s^2 + 6s + 2K(s + 2)},\quad \text{system stable}\right]$$

(C.E.I. Part 2, Control Systems, 1972)

4. A negative-feedback amplifier consists of three RC coupled stages whose midband voltage gains are 20, 15 and 10 respectively. The fraction of the output voltage fed back in antiphase to the input voltage is 1/60. If the r.m.s. input voltage is of 0.3 V magnitude at the midband frequency, calculate the r.m.s. output voltage at this frequency.

$$\left[\frac{V_o}{V_i} = 58.8,\ 17.64\ \text{V}\right]$$

5. A rotor having a moment of inertia $J(\text{kg m}^2)$ is coupled to a viscous damper requiring a torque of $F(\text{N m rad}^{-1}\text{ s})$. Draw a block diagram relating the rotor angular velocity $\omega(\text{rad s}^{-1})$ to the applied torque $T(\text{N m})$. Hence obtain the transfer function relating these variables.

$$\left[\frac{\omega}{T}(s) = \frac{1}{F + sJ}\right]$$

6. A control system has the configuration shown in fig. 2.23, where K_1 and K_2 are gain constants in the forward and feedback loops respectively. It is desired that the forward transfer function of the system should be

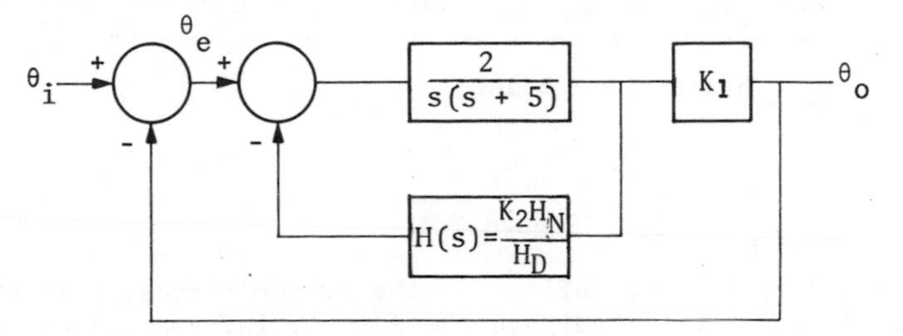

Figure 2.23

$$\frac{\theta_o}{\theta_e}(s) = \frac{100(s + 10)}{s(s + 5)(s + 20)}$$

Determine the values of K_1 and K_2 and the transfer function of the feedback loop $H(s)$ in order that this forward transfer function may be obtained.

$$\left[K_1 = 50,\ K_2 = 5,\ H(s) = \frac{s(s + 5)}{s + 10}\right]$$

7. Determine θ_o for the system shown in fig. 2.24.

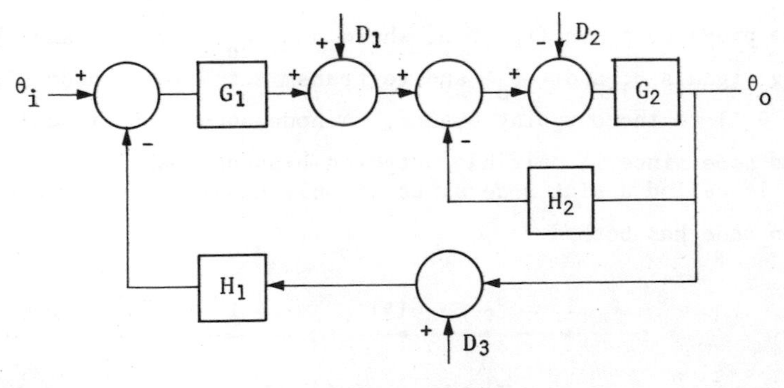

Figure 2.24

$$\left[\theta_o = \frac{G_1G_2\theta_i + G_2D_1 - G_2D_2 - G_1G_2H_1D_3}{1 + G_2H_2 + G_1G_2H_1}\right]$$

3 SIGNAL FLOW GRAPHS

Signal flow graphs consist of a network in which *nodes* representing each of the system variables are connected by directed *branches*, see fig. 3.1. A branch acts as a one-way signal multiplier, the ratio

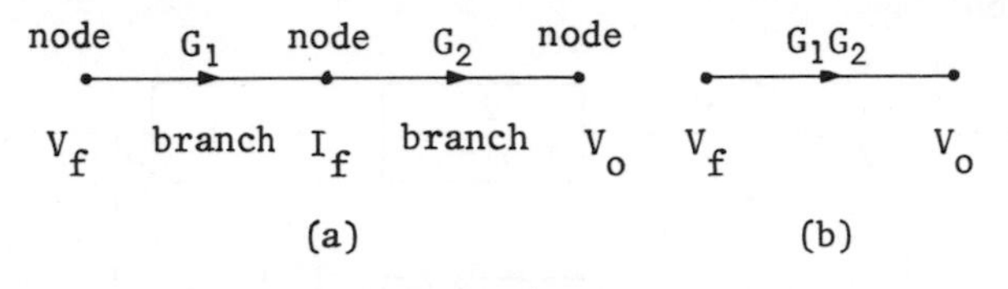

Figure 3.1

of output to input being defined as the *transmittance*, with the arrowhead being used to indicate the flow of information as in a block diagram.

For the general control system as shown in fig. 3.2, a flow diagram

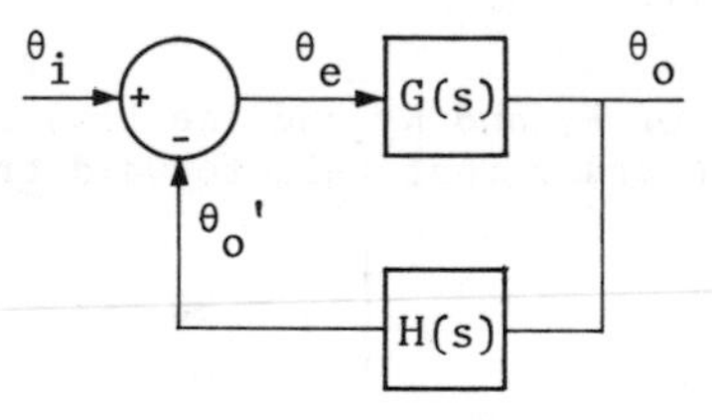

Figure 3.2

may be prepared as in fig. 3.3, where node θ_e serves to add the incoming signals θ_i and $-\theta_o'$ and to transmit the total node signal $(\theta_1 - \theta_o')$ to the outgoing branch. A node such as θ_i is called a *source node* since it only has outgoing branches, while a node such as θ_o is called a *sink node* since it only has incoming branches. A *common node* has both.

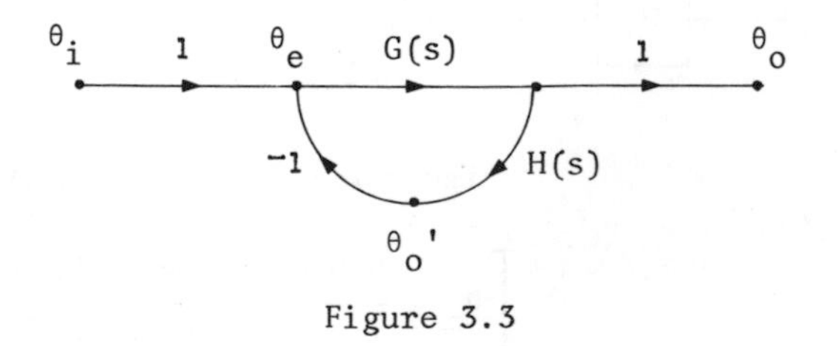

Figure 3.3

A *path* is any connected sequence of branches whose arrows are in the same direction, such as θ_i to θ_o in fig. 3.3.

Mason's Rule

This is a method that can be used when determining an overall transfer function from a signal flow diagram. It is given by

$$T = \frac{1}{\Delta} \sum_m T_m \Delta_m \tag{3.1}$$

where T_m is the transmittance of the mth forward path between the source node and the sink node and Δ is the determinant of the graph and is given by

$$\Delta = 1 - \sum T_{m_1} + \sum T_{m_2} - \sum T_{m_3} + \ldots \tag{3.2}$$

where T_{m_1} is the transmittance of each closed path, T_{m_2} is the product of the transmittances of two non-touching loops (i.e. having no nodes in common) and T_{m_3} is the product of the transmittances of three non-touching loops. Δ_m is the co-factor of the forward path m and is the value of Δ for that part of the graph that does *not* touch the path m.

Example 3.1

Obtain the signal flow graph for the block diagram of fig. 3.4 and obtain the closed-loop transfer function by using Mason's rule.

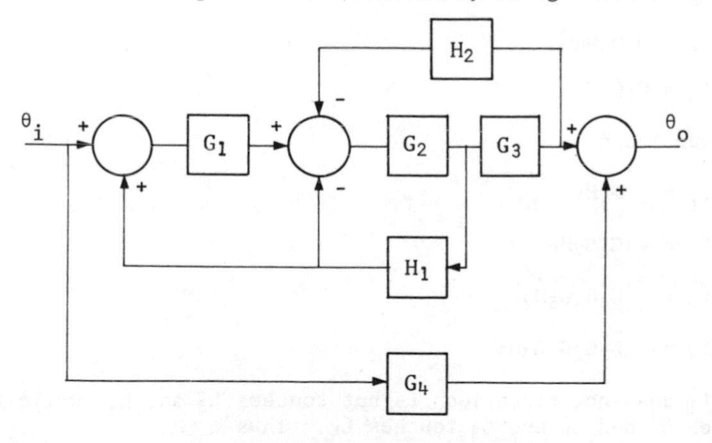

Figure 3.4

There are two forward paths $T_1 = G_1G_2G_3$ and $T_2 = G_4$ and three feedback or closed-loop paths, $L_1 = -G_2H_1$, $L_2 = G_1G_2H_1$ and $L_3 = -G_2G_3H_2$. All loops touch one another, hence

$$\Delta = 1 - (L_1 + L_2 + L_3) = 1 + G_2H_1 - G_1G_2H_1 + G_2G_3H_2$$

Removing loops that touch T_1 yields $\Delta_1 = 1$. But no loops touch path T_2, thus $\Delta_2 = \Delta$.

$$\frac{\theta_o}{\theta_i} = \frac{T_1\Delta_1 + T_2\Delta_2}{\Delta}$$

$$= \frac{G_1G_2G_3 + G_4(1 + G_2H_1 - G_1G_2H_1 + G_2G_3H_2)}{1 + G_2H_1 - G_1G_2H_1 + G_2G_3H_2} \tag{3.3}$$

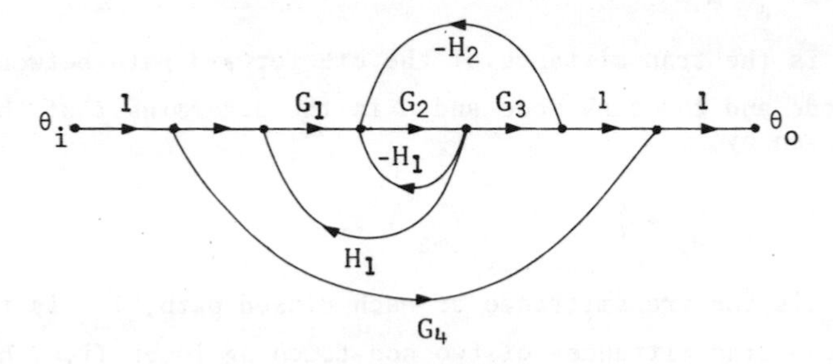

Figure 3.5

Example 3.2

Obtain the closed-loop transfer functions for the system shown in fig. 3.6.

There are three forward paths

$$T_1 = G_1G_2G_3G_4G_5$$

$$T_2 = G_1G_6G_4G_5$$

$$T_3 = G_1G_2G_7$$

and four loops

$$L_1 = - G_4H_1$$

$$L_2 = - G_2G_7H_2$$

$$L_3 = - G_6G_4G_5H_2$$

$$L_4 = - G_2G_3G_4G_5H_2$$

loop L_1 does not touch loop L_2 but touches L_3 and L_4, while L_2 touches L_3 and L_4 and L_3 touches L_4. Thus

$$\Delta = 1 - (L_1 + L_2 + L_3 + L_4) + L_1L_2$$

By removing the loops that touch T_1, the cofactor $\Delta_1 = 1$; similarly $\Delta_2 = 1$. But loop L_1 does not touch T_3, so that

$$\Delta_3 = 1 - L_1$$

$$\frac{\theta_o}{\theta_i} = \frac{1}{\Delta}(T_1\Delta_1 + T_2\Delta_2 + T_3\Delta_3) \tag{3.4}$$

$$\frac{\theta_o}{\theta_i} = \frac{G_1G_4G_5(G_2G_3 + G_6) + G_1G_2G_7(1 + G_4H_1)}{1 + G_2G_7H_2 + G_4H_1(1 + G_2G_7H_2) + G_4G_5H_2(G_2G_3 + G_6)} \tag{3.5}$$

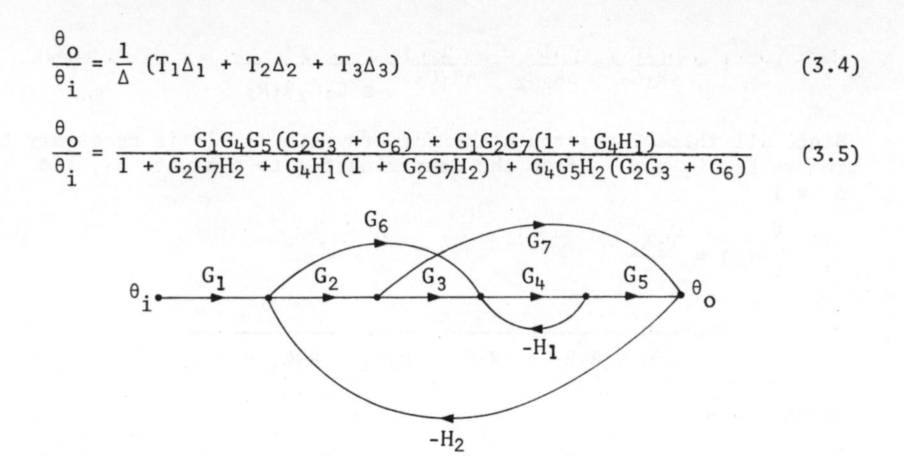

Figure 3.6

Example 3.3

Obtain the closed-loop transfer function $V_o/V_i(s)$ in the system shown in fig. 3.7.

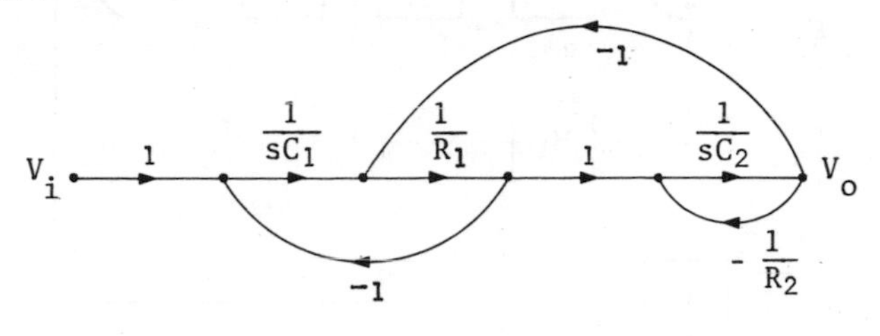

Figure 3.7

There is only one forward path

$$T_1 = \frac{1}{s^2C_1C_2R_1}$$

and three loops

$$L_1 = -\frac{1}{sC_1R_1}$$

$$L_2 = -\frac{1}{sC_2R_2}$$

$$L_3 = -\frac{1}{sC_2R_1}$$

Loop L_1 does not touch loop L_2, while L_1 touches L_3 and L_2 touches L_3. Therefore

$$\Delta = 1 - (L_1 + L_2 + L_3) + L_1L_2$$

$$\Delta = 1 + \frac{1}{sR_1C_1} + \frac{1}{sR_2C_2} + \frac{1}{sR_1C_2} + \frac{1}{s^2C_1C_2R_1R_2}$$

Since all three loops touch the forward path T_1, it is necessary to remove L_1, L_2 and L_3 from the determinant Δ to evaluate Δ_1, i.e. $\Delta_1 = 1$.

$$\frac{V_o}{V_i}(s) = \frac{T_1\Delta_1}{\Delta}$$

$$= \frac{R_2}{s^2C_1C_2R_1R_2 + (R_1C_1 + R_2C_2 + R_2C_1)s + 1} \qquad (3.6)$$

Example 3.4

A control system is shown in fig. 3.8. Prepare (a) a signal flow diagram and (b) a simplified block diagram.

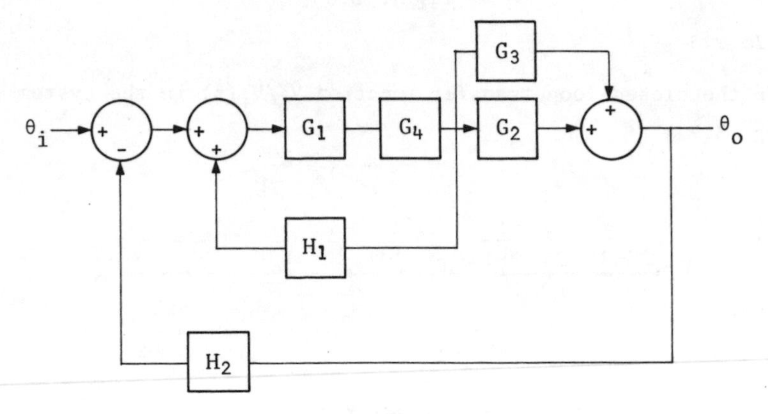

Figure 3.8

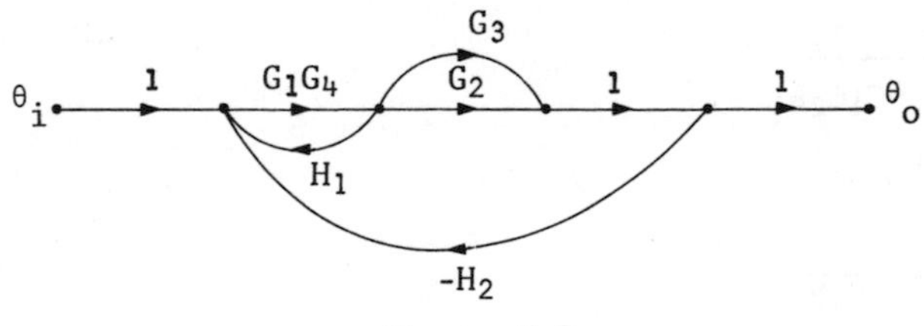

Figure 3.9

Two paths

$T_1 = G_1G_2G_4$

$T_2 = G_1G_3G_4$

Three loops

$L_1 = G_1G_4H_1$

$$L_2 = -G_1G_2G_4H_2$$

$$L_3 = -G_1G_3G_4H_2$$

There are no non-touching loops and all the loops touch both forward paths, therefore

$$\Delta = 1 - (L_1 + L_2 + L_3)$$

$$\Delta_1 = 1$$

$$\Delta_2 = 1$$

$$\frac{\theta_o}{\theta_i} = \frac{T_1\Delta_1 + T_2\Delta_2}{\Delta}$$

$$= \frac{G_1G_2G_4 + G_1G_3G_4}{1 - G_1G_4H_1 + G_1G_2G_4H_2 + G_1G_3G_4H_2} \qquad (3.7)$$

To obtain a simplified block diagram compare eqn 3.7 with the standard form of closed-loop transfer function $G/(1 + GH)$, i.e. let $G = G_1G_4(G_2 + G_3)$.

$$GH = G_1G_4(G_2H_2 + G_3H_2 - H_1)$$

therefore

$$H = \frac{G_2H_2 + G_3H_2 - H_1}{G_2 + G_3} \qquad (3.8)$$

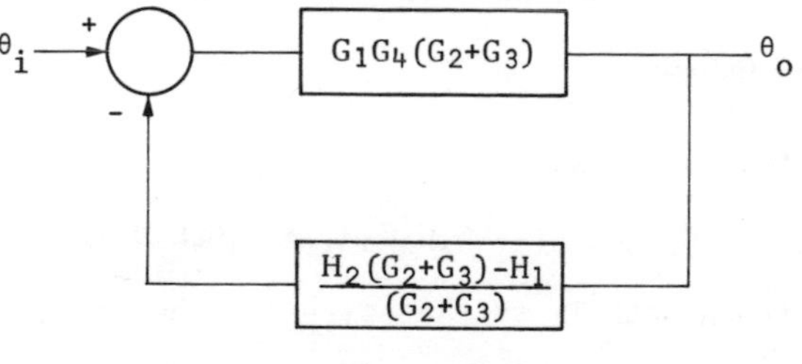

Figure 3.10

Example 3.5

Derive the general expression relating the output velocity ω_o to the load torque T_L and input signal ω_i from the signal flow diagram shown in fig. 3.11, when

$$G_1 = \frac{90}{1 + 0.06s}$$

$$G_2 = \frac{2.5}{1 + 0.19s}$$

$G_3 = 25$

$G_4 = 3.8$

$G_5 = \dfrac{1}{1355s}$

$G_6 = 0.4$

$G_7 = -\ 0.4$

$H_1 = -\ 0.001$

$H_2 = -\ 4$

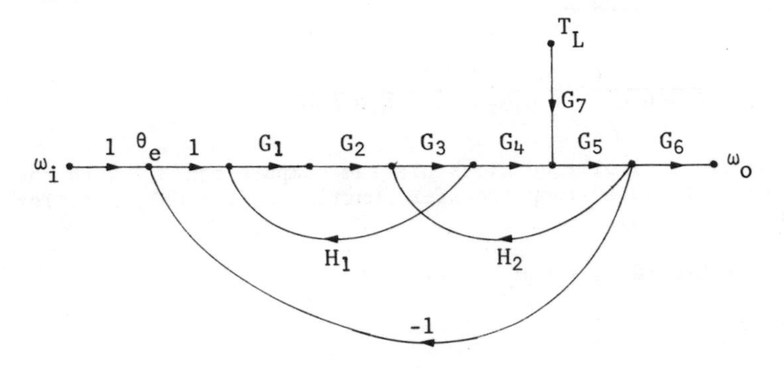

Figure 3.11

The forward transmittance path between ω_i and ω_o is

$$T_1 = G_1G_2G_3G_4G_5G_6$$

The diagram determinant is

$$\Delta = 1 - [(-\ G_1G_2G_3G_4G_5) + G_1G_2G_3H_1 + G_3G_4G_5H_2]$$

All the loops touch path T_1 so that $\Delta_1 = 1$. Therefore

$$\frac{\omega_o}{\omega_i} = \frac{G_1G_2G_3G_4G_5G_6}{1 + G_1G_2G_3G_4G_5 - G_1G_2G_3H_1 - G_3G_4G_5H_2}$$

The relationship between T_L and ω_o is found in a similar way: $T_2 = G_5G_6G_7$, $\Delta_1 = 1 - G_1G_2G_3H_1$. Thus

$$\frac{\omega_o}{T_L} = \frac{G_5G_6G_7(1 - G_1G_2G_3H_1)}{1 + G_1G_2G_3G_4G_5 - G_1G_2G_3G_1 - G_3G_4G_5H_2} \tag{3.9}$$

On substituting the given values and using superposition

$$\omega_o = \frac{22.5\omega_i - 2.36 \times 10^{-3}[1 + 0.18(1 + 0.06s)(1 + 0.19s)]T_L}{(1 + 0.06s)(1 + 0.19s)(1 + 3.57s) + 56(1 + 0.36s)} \tag{3.10}$$

Problems

1. Fig. 3.12 shows the signal flow graph for a speed-control system; M(s) is the Laplace transform of the load torque. Find the transfer function $\theta_o(s)/M(s)$ and show that the steady-state decrease in the controlled variable due to unit step of applied load torque is approximately $1/K_1K_3$, provided that $K_1K_5 \gg K_2$ and $K_1K_3K_4K_5 \gg 1$.

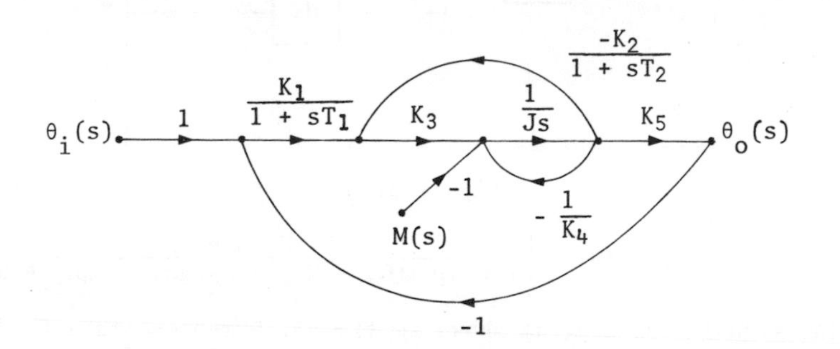

Figure 3.12

(C.E.I. S & C Eng., Dec 1966)

$$\left[- \frac{K_5}{sJ + \frac{K_1K_3K_5}{1 + sT_1} + \frac{K_2K_3}{1 + sT_2} + \frac{1}{K_4}} \right]$$

2. What are the advantages of signal flow graphs over block diagrams? Describe the functions of nodes, branches and arrows.

Draw a block diagram for the signal flow graph shown in fig. 3.13 and find the transfer function between θ_o and θ_i.

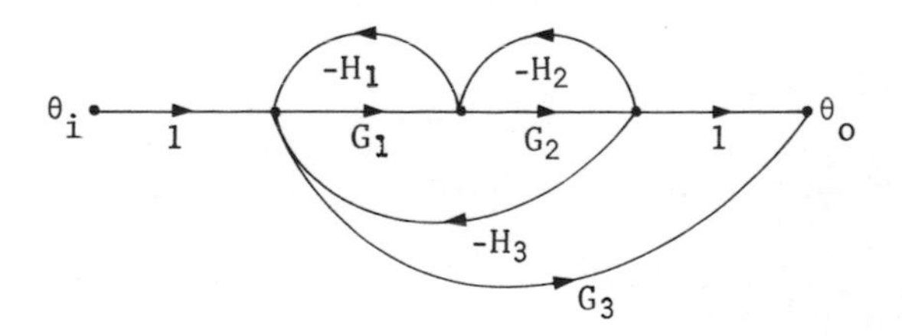

Figure 3.13

(C.E.I. S & C Eng., Oct 1968)

$$\left[\frac{G_1G_2 + G_3(1 + G_2H_2)}{1 + G_1H_1 + G_2H_2 + G_1G_2H_3} \right]$$

3. Draw a signal flow graph for the control system described by the block diagram in fig. 3.14. From the signal flow graph determine the relationship between O and I.

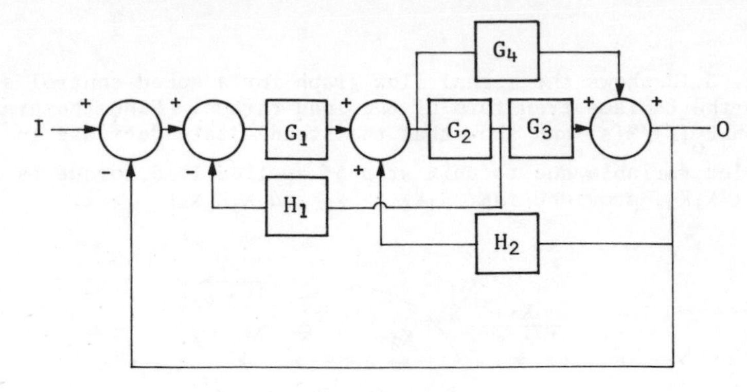

Figure 3.14

(C.E.I. S & C Eng., 1970) $\left[\dfrac{G_1(G_2G_3 + G_4)}{1 + G_1G_2(G_3 + H_1) - G_2G_3H_2 - G_4H_2 + G_1G_4}\right]$

4. The signal flow diagram shown in fig. 3.15 represents a feedback system in which the main feedback path is AA'. Determine the conditions for the system to be stable (a) with AA' open and (b) with AA' closed.

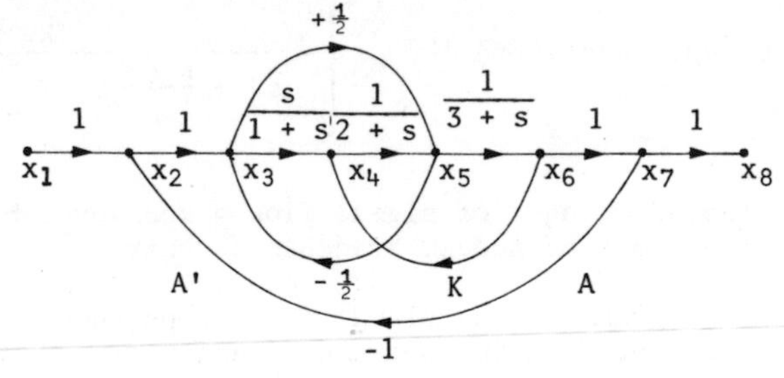

Figure 3.15

(C.E.I. E.F.C., 1972) [K < 7.5; K < 8.5]

4 STEADY-STATE AND TRANSIENT RESPONSE

The solution of the differential equation of most physical systems is in two parts. The transient solution or natural response is that part of the total response that approaches zero as time approaches infinity (complementary function), while the steady-state solution or forced response is that part of the total response that *does not* approach zero as time approaches infinity (particular integral).

It has already been mentioned in chapter 1 that the differential equation for some systems can become extremely cumbersome to handle in the traditional manner and so, as in the case of transfer functions, the Laplace operator s is introduced to solve the linear system equations.

In order to analyse and design control systems, taking into account both the transient and the steady-state behaviour, a number of different input signals must be applied to the system. In practice the input signals will vary in complexity and in random fashion with respect to time. It is therefore necessary to use standard test signals in order to study the performance and reaction of the system.

The most commonly used test signals are those of step, ramp, acceleration, impulse and sinusoidal functions. Which of these typical signals to use when analysing system characteristics, is usually determined by the form of the input that the system will be subjected to under normal operating conditions. If a system is subjected to sudden disturbances, a step function of time would be a good test signal, while if the input is a gradually changing function of time, then a ramp function may be a good test signal.

Example 4.1

A flywheel driven by an electric motor is automatically controlled to follow the movement of a hand-wheel. The inclusive moment of inertia of the flywheel is 150 kg m^2 and the motor torque applied to it is 2400 N m per radian of misalignment between the flywheel and the hand-wheel. The viscous friction is equivalent to a torque of 600 N m rad^{-1} s.

If the hand-wheel is suddenly turned through 60° when the system is at rest, determine an expression for the subsequent angular position of the flywheel in relation to time.

Since the motor torque applied to the flywheel must equal the reaction torque

$$150s^2\theta_o + 600s\theta_o = 2400\ (\theta_i - \theta_o)$$

Therefore

$$\frac{\theta_o}{\theta_i}(s) = \frac{2400}{150s^2 + 600s + 2400}$$

For a step of 60°, $\theta_i(s) = \pi/(3s)$ and the response is then

$$\theta_o(s) = \frac{16\pi/3}{s(s^2 + 4s + 16)}$$

The equivalent partial fractions are

$$\frac{A}{s} + \frac{Bs + C}{(s + 2)^2 + 12}$$

where $A = \pi/3 = - B$ and $C = - 4\pi/3$. Therefore

$$\theta_o(s) = \frac{\pi/3}{s} - \frac{(\pi/3)s + (4\pi/3)}{(s + 2)^2 + 12}$$

and from transform tables

$$\theta_o(t) = \frac{\pi}{3}\left\{1 - e^{-2t} \cos \sqrt{12}t - \frac{2}{\sqrt{12}} e^{-2t} \sin \sqrt{12}t\right\}$$

or $\quad \theta_o(t) = \frac{\pi}{3} [1 - e^{-2t}(\cos 3.46t + 0.58 \sin 3.46t)] \qquad (4.1)$

Example 4.2

The speed of a flywheel, driven by an electric motor, is to be controlled from the setting of an input potentiometer using a closed-loop automatic speed-control system. The inclusive moment of inertia of the flywheel and motor is 100 kg m^2 and a speed error of 1 rad s^{-1} produces a torque on the flywheel of 45 N m. Frictional torque is 5 N m when the flywheel velocity is 1 rad s^{-1}. With the system at rest, the input potentiometer setting is suddenly increased from zero to 50 rev/min. Derive the relationship between the subsequent flywheel velocity and time, and calculate the steady-state velocity error of the flywheel.

The differential equation for a speed-control system may be written as

$$J \frac{d\omega_o}{dt} + F\omega_o = K(\omega_i - \omega_o)$$

In Laplace form, this equation becomes

$$100s\omega_o + (5 + 45)\omega_o = \frac{45 \times 50 \times 2\pi}{60s}$$

therefore

$$\omega_o(s) = \frac{0.75\pi}{s(s + 0.5)}$$

and from transform tables

$$\omega_o(t) = 1.5\pi(1 - e^{-0.5t}) \text{ rad s}^{-1} \tag{4.2}$$

The steady-state velocity is 1.5π rad s^{-1} or 45 rev/min, therefore the steady-state error is 5 rev/min.

Example 4.3

A simple proportional position-control system has a total effective moment of inertia, referred to the output shaft, of 2000 kg m^2, while the controller gain, in terms of torque output per minute of misalignment, is 2200 N m. If the system is critically damped by means of output derivative feedback, determine the steady-state positional error when the input shaft is rotating at a speed of 3 rev/min. If the derivative of error control is added to the system incorporating a time constant of 0.1 s and the output derivative feedback is adjusted to give critical damping, determine the new steady-state error if the input conditions remain unaltered.

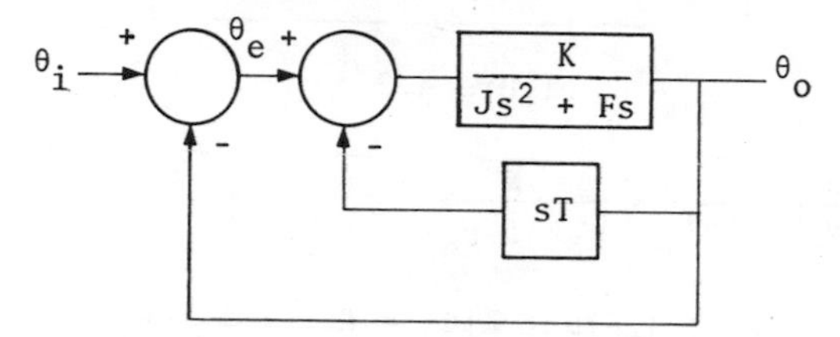

Figure 4.1

The idea of introducing derivative feedback is to improve the transient performance of a system. (Students should check this fact out for themselves.)

From fig. 4.1

$$\frac{\theta_o}{\theta_e - Ts\theta_o} = \frac{K}{Js^2 + Fs}$$

therefore

$$[Js^2 + (F + TK)s + K]\theta_o = K\theta_i$$

For critical damping $F + TK = 2\sqrt{(JK)}$.

$$K = 2200 \times 60 \times \frac{180}{\pi} = 755 \times 10^4 \text{ N m rad}^{-1}$$

$$\text{Input speed } \omega = 3 \times \frac{2\pi}{60} = \frac{\pi}{10} \text{ rad s}^{-1}$$

therefore

$$(F + TK) = 2\surd(2000 \times 755 \times 10^4) = 246 \times 10^3$$

For steady-state conditions

$$\theta_e = \frac{246 \times 10^3 \times \pi}{755 \times 10^4 \times 10} = 0.01 \text{ rad}$$

With derivative of error added to the system, the transient performance could still be improved but, included with derivative feedback signals, may worsen the velocity lag error condition.

The system is now as shown in fig. 4.2.

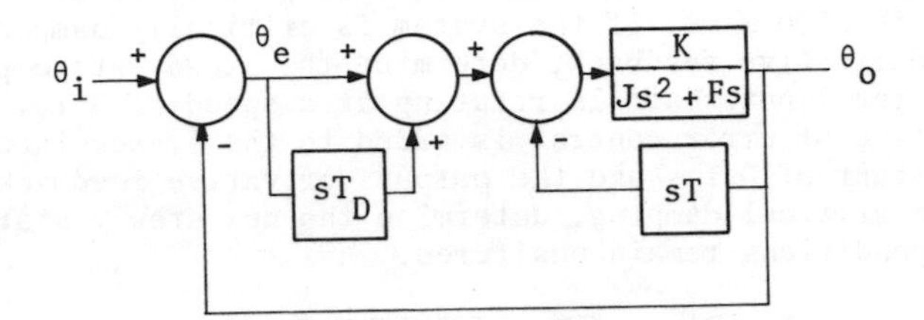

Figure 4.2

$$\frac{\theta_o}{\theta_e + T_D s\theta_e - Ts\theta_o} = \frac{K}{Js^2 + Fs}$$

$$[Js^2 + (F + KT + T_D K)s + K]\theta_o = K\theta_i + T_D Ks\theta_i$$

This system is still to be critically damped, so that

$$F + KT + T_D K = 2\surd(JK) = 246 \times 10^3$$

For the steady-state conditions, with $T_D = 0.1$ s

$$\theta_e = \omega \frac{(246 \times 10^3 - 755 \times 10^3)}{755 \times 10^4}$$

$$\theta_e = -\frac{509 \times 10^3}{755 \times 10^4} \times \frac{\pi}{10} = -0.02$$

(A good exercise for the student would be to examine this problem for critical damping when $T = 0$ and $T_D = 0.1$ and, say $T = 0.2$ when $T_D = 0$.)

Example 4.4

The angular position of an aerial is controlled by a closed-loop automatic-control system to follow an input wheel. The input wheel is maintained in sinusoidal oscillation through ± 34° with an angular frequency $\omega = 1$ rad s^{-1}. The moving part of the aerial system has a moment of inertia of 200 kg m^2 and a viscous frictional torque of 1600 N m rad^{-1} s.

Calculate the amplitude of swing of the aerial and the time lag between the aerial and the input wheel, if the system is critically damped.

$$(Js^2 + Fs)\theta_o = K(\theta_i - \theta_o)$$

for critical damping, $F = 2\sqrt{(JK)}$, therefore

$$K = \frac{1600^2}{4 \times 200} = 3200 \text{ N m rad}^{-1}$$

Natural undamped frequency is

$$\omega_n = \sqrt{\frac{3200}{200}} = 4 \text{ rad s}^{-1}$$

In the alternative form, the torque equation may be expressed as

$$(s^2 + 2\zeta\omega_n s + \omega_n^2)\theta_o = \omega_n^2\theta_i$$

and for a sinusoidal input, $\theta_i = \theta \sin \omega t$, the steady-state output is found to be

$$\frac{\theta}{\left[\left(\frac{\omega^2}{\omega_n^2} - 1\right)^2 + 4\zeta^2 \frac{\omega^2}{\omega_n^2}\right]^{\frac{1}{2}}}$$

Thus the amplitude of aerial swing is

$$\frac{34}{\left[\left(\frac{1}{16} - 1\right)^2 + \frac{4 \times 1}{16}\right]^{\frac{1}{2}}} \quad \text{or } 32°$$

The output lags the input by an angle given by $\tan^{-1}[2\omega\omega_n\zeta/(\omega_n^2 - \omega^2)]$. In this case the angle is

$$\tan^{-1} \frac{2 \times 1 \times 4 \times 1}{16 - 1} \quad \text{or } 28°$$

Therefore the time lag is 0.49 seconds.

Example 4.5

A three-term controller is described by the equation

$$\theta_c(t) = 20\left[e(t) + \frac{1}{T_r}\int_0^t e(t)\,dt + T_d \frac{de}{dt}(t)\right]$$

where $e(t)$ = system error

$\theta_c(t)$ = controller output

T_r = reset time

T_d = derivative time

This is used to control a process with transfer function

$$G(s) = \frac{40}{10s^2 + 80s + 800}$$

Unity feedback is used.

(a) If integral action is not employed, find the derivative time required to make the closed-loop damping ratio unity.

(b) If this value of derivative time is maintained, determine the minimum value of reset time that can be used without instability arising.

From the given data

$$\frac{\theta_o}{e}(s) = \frac{800[1 + (1/sT_r) + sT_d]}{10s^2 + 80s + 800}$$

(a) If $T_r = 0$

$$\frac{\theta_o}{e}(s) = \frac{80[1 + sT_d]}{s^2 + 8s + 80}$$

$$\frac{\theta_o}{\theta_i}(s) = \frac{80[1 + sT_d]}{s^2 + s(8 + 80T_d) + 160}$$

compare characteristic equation with $s^2 + 2\zeta\omega_n s + \omega_n^2$, therefore

$$\omega_n^2 = 160$$

$$\omega_n = 12.65 \text{ rad s}^{-1}$$

$$2\zeta\omega_n = 8 + 80T_d$$

for critical damping $\zeta = 1$, therefore

$$T_d = \frac{25.3 - 8}{80} = 0.216 \text{ seconds}$$

(b) $$\frac{\theta_o}{e}(s) = \frac{80[1 + (1/sT_r) + 0.216s]}{s^2 + 8s + 80}$$

$$\frac{\theta_o}{\theta_i}(s) = \frac{80[1 + (1/sT_r) + 0.216s]}{s^2 + 8s + 160 + (80/sT_r) + 17.3s}$$

The characteristic equation is

$$s^3 + 25.3s^2 + 160s + \frac{80}{T_r} = 0$$

Applying Routh's criterion, the array is

$$\begin{array}{ll} 1 & 160 \\ 25.3 & 80/T_r \\ \dfrac{4048 - (80/T_r)}{25.3} & \\ 80/T_r & \end{array}$$

For critical stability

$$4048 = \frac{80}{T_r} \quad \text{or} \quad T_r = 0.01976 \text{ s}$$

(C.E.I. S & C Eng., 1971)

Example 4.6

Fig. 4.3a shows a mechanical vibratory system. When 8.9 N of force is applied to the system, the mass oscillates, as shown in fig. 4.3b. Determine M, F and K of the system from this response curve.

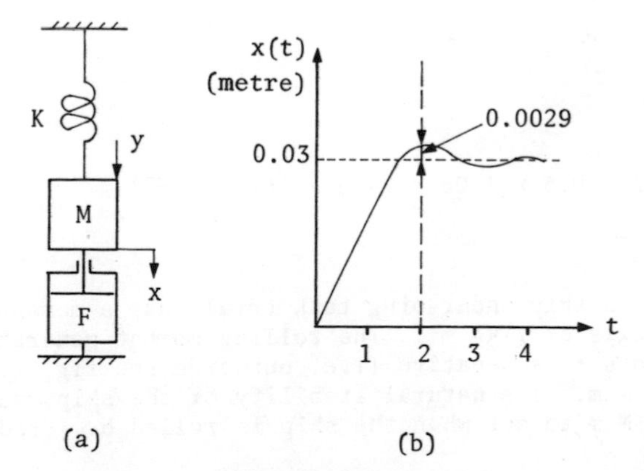

Figure 4.3

The transfer function of this system is

$$\frac{x}{y}(s) = \frac{1}{Ms^2 + Fs + K}$$

$$y(s) = \frac{8.9}{s}$$

$$x(s) = \frac{8.9}{s(Ms^2 + Fs + K)}$$

The steady-state value of x is

$$x = \lim_{s \to 0} sx(s) = \frac{8.9}{K} = 0.03$$

therefore

$$K = \frac{8.9}{0.03} = 297 \text{ N m}^{-1}$$

$M_{max} = 9.66\%$ corresponds to $\zeta = 0.6$. The corresponding time is

$$t_{max} = \frac{\pi}{\omega_n \sqrt{(1 - \zeta^2)}} = \frac{\pi}{0.8\omega_n} = 2 \text{ s}$$

therefore

$$\omega_n = \frac{3.14}{0.8 \times 2} = 1.96 \text{ rad s}^{-1}$$

since $\omega_n^2 = K/M$

$$M = \frac{297}{1.96^2} = 77.3 \text{ kg}$$

also $2\zeta\omega_n = \frac{F}{M}$

therefore

$$F = 2 \times 0.6 \times 1.96 \times 77.3 = 181.8 \text{ N m}^{-1} \text{ s}$$

Example 4.7

A model of a ship undergoing tank trials has a moment of inertia about the roll axis of 1 kg m^2. The rolling moment generated by a rate of turn r rad s^{-1} is negative (i.e. outwards see fig. 4.4) and of magnitude 0.8r N m. The natural stability of the ship causes a restoring moment 2ϕ N m to act when the ship is rolled by ϕ rad.

It is decided to fit roll-control fins, which produce a positive (clockwise looking forward) moment of 4δ N m, where δ rad is the deflection angle. If the roll-control system drives the fins according to the law

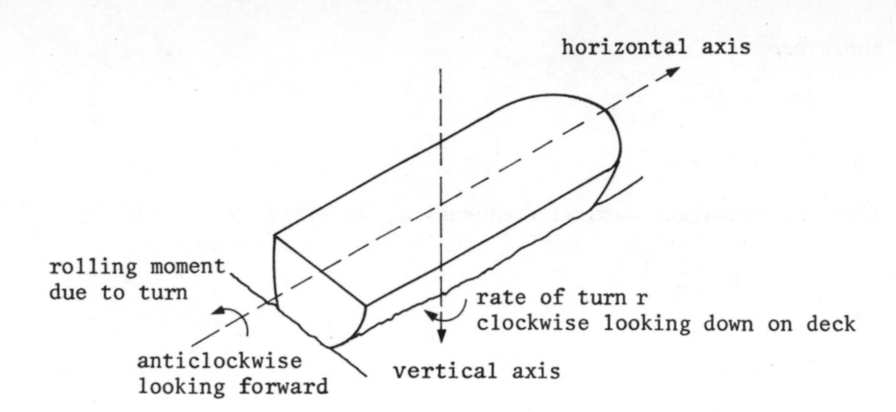

Figure 4.4

$$\delta = a\phi + b\frac{d\phi}{dt} + cr$$

find a, b and c such that (a) the ship remains upright (unrolled) during a steady turn and (b) the roll transients have a natural frequency of 2 rad s^{-1} and a damping ratio of 0.5.

(C.E.I. C.S.E., 1975)

This problem is one of balancing moments against the inertia moment of the boat

Restoring moment = 2ϕ

Roll-control moment = $4\left(a\phi + b\frac{d\phi}{dt} + cr\right)$

Rolling moment = $0.8r$

For rolling

$$0.8r - 4\left(a\phi + b\frac{d\phi}{dt} + cr\right) - 2\phi = J\frac{d^2\phi}{dt^2} = \frac{d^2\phi}{dt^2}$$

Rearrange this equation using the Laplace operator.

$$[s^2 + 4bs + (4a + 2)]\phi = (0.8 - 4c)\ r$$

For steady state

$$(4a + 2)\phi = (0.8 - 4c)\ R$$

If $\phi = 0$

$$c = 0.2$$

Compare the quadratic equation above with the basic equation

$$(s^2 + 2\zeta\omega_n s + \omega_n^{\ 2})\phi = 0$$

therefore

$$2\zeta\omega_n = 4b$$

and $(4a + 2) = \omega_n^2$

Also the transient natural frequency ω_D is given by

$$\omega_D = \omega_n\sqrt{(1 - \zeta^2)}$$

therefore

$$4 = \omega_n^2(1 - 0.25)$$

or $\omega_n^2 = \frac{16}{3}$

thus $4a + 2 = \frac{16}{3}$

or $a = \frac{5}{6}$

thus $4b = 2 \times 0.5 \times \frac{4}{\sqrt{3}}$

or $b = \frac{1}{\sqrt{3}}$

Problems

1. A servo system for the positional control of a rotatable mass is stabilised by viscous friction damping which is three-quarters of that needed for critical damping. The undamped frequency of the system is 12 Hz. Derive an expression for the output of the system if the input control is suddenly moved to a new position, the system being initially at rest. Hence find the maximum overshoot.

$$\left\{\theta_o(t) = \theta_i\left[1 - 1.5\ e^{-56.6t}\ \sin\ (50t + 0.72)\right];\ 2.8\%\right\}$$

2. A servo mechanism, designed to control the angular position of a rotatable mass, is stabilised by means of acceleration feedback. The moment of inertia of the system is 10^{-5} kg m^2, the viscous frictional torque per radian per second is 10^{-4} N m and the motor torque T_m is given by

$$T_m = 4 \times 10^{-3}\left[\theta_e + ks^2\theta_o\right]\ \text{N m}$$

Draw the block diagram of the system and develop the control equation. Hence determine the value of k in order that the damping shall be critical. What is the steady-state error for an input signal of 1.26 rad s^{-1}?

[2.34×10^{-3}; 0.0314 rad]

3. A control system of inertia 160 kg m^2 is operated with viscous friction damping only; the value of this damping coefficient is 640 N m rad^{-1} s. If the system has a damping ratio of 0.4, determine the natural frequency and the steady-state error when subjected to a ramp function input of 10 rev/min.

Derivative error control is added to the system and the gain adjusted such that the natural frequency is raised to four times its original value, while the damping ratio is maintained at its original value of 0.4. Determine the new steady-state error for the same ramp-function input and estimate the value of the derivative time constant introduced in the system.

[5 rad s^{-1}; 0.168 rad; 0.0105 rad; 0.03 s]

4. The angular position of a flywheel on a test rig is controlled by an error-actuated closed-loop automatic control system to follow the motion of an input lever. The lever is maintained in sinusoidal oscillations through ± 60° with an angular frequency $\omega = 2$ rad s^{-1}. The inclusive moment of inertia of the flywheel is 150 kg m^2 and the stiffness of the control is 2400 N m per radian of misalignment. Calculate the viscous frictional torque required to produce critical damping.

Assuming critical damping, calculate the amplitude of swing of the flywheel and the time lag between the flywheel and the control lever.

[1200 N m rad^{-1} s; 48°; 0.464 s]

5. The outflow of liquid from a vessel 0.5 m in diameter is proportional to the head of water in the vessel. A ball valve operated by the liquid level regulates the inflow to $0.125x$ m^3 s^{-1}, where x is the fall in level below the desired value of 1.5 m. If the stop valve in the outflow pipe is suddenly opened, so that the outflow changes from zero to $0.01h$ m^3 s^{-1}, where h is the actual level of the vessel, determine

(a) the steady-state value of the liquid level in the tank
(b) the time taken for the level in the tank to fall 8 cm
(c) the instantaneous values of the inflow and outflow, t seconds after the opening of the valve.

[1.388 m; 1.845 s; $0.0139(1 - e^{-0.6875t})$;

$0.0139(1 + 0.079e^{-0.6875t})$]

6. The forward path of a control loop is described by the differential equation

$$\frac{1}{\omega_n^2}\frac{d^2\theta_o}{dt^2} + \frac{2c}{\omega_n}\frac{d\theta_o}{dt} + \theta_o = K_c\theta_e$$

The gain of the feedback is denoted by K_f. If the parameters of the system are

Inertia = 322 lb ft^2 or 10 kg m^2

Stiffness = 160 ft lbf rad^{-1} or 160 N m rad^{-1}

Gain K_c = 2.0

Damping ratio = 0.8

find the value of K_f such that the damping ratio of the closed-loop system is 0.5. Determine an expression for the transient response to a unit step input, assuming that at t = 0 both θ_o and $d\theta_o/dt$ are zero.

[$K_f = 0.78$, $0.78 - e^{-3.2t}(0.78 \cos 5.5t + 0.45 \sin 5.5t)$]

(C.E.I. S & C Eng., 1968)

7. The load characteristic of a computing element consists entirely of inertia and viscous damping. The element is driven by a simple position-control servo mechanism where the output torque of the servo motor is $K(1 + T_D s)\theta_e$, where K is the proportional gain constant, T_D is the derivative action time and θ_e is the error.

With the proportional and derivative controls set to 50 per cent of their maximum values, the system is tested and it is found that a step displacement input produces a damped oscillatory response of frequency 40 rad s^{-1}, the ratio of successive amplitudes measured on the same side of the settling position being 0.12. With a constant-velocity input of 60 rev/min, the steady-state velocity lag is 2°.

If the control is to be adjusted so that the steady-state lag is $1\frac{1}{4}$° with a constant-velocity input of 60 rev/min, and the overall system damping is to be half of that required for critical damping, determine the settings of the proportional and derivative controls.

[P.C.S. = 77% of maximum setting;
D.C.S. = 84.5% of maximum setting]

5 NYQUIST ANALYSIS AND STABILITY

The analysis of control systems with feedback may also be carried out using a method based on frequency response. The essence of this method, often referred to as a Nyquist plot, is a graphical procedure for determining absolute and relative stability of closed-loop control systems. Information about stability is available directly from a polar plot of the sinusoidal *open-loop* transfer function $G(j\omega)H(j\omega)$, once the feedback system has been reduced to standard form.

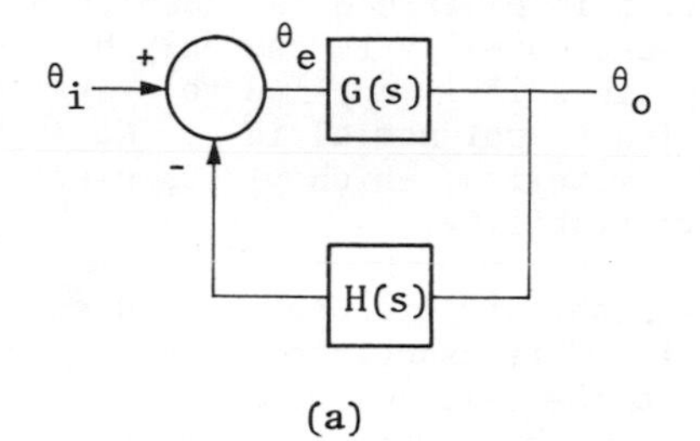

(a)

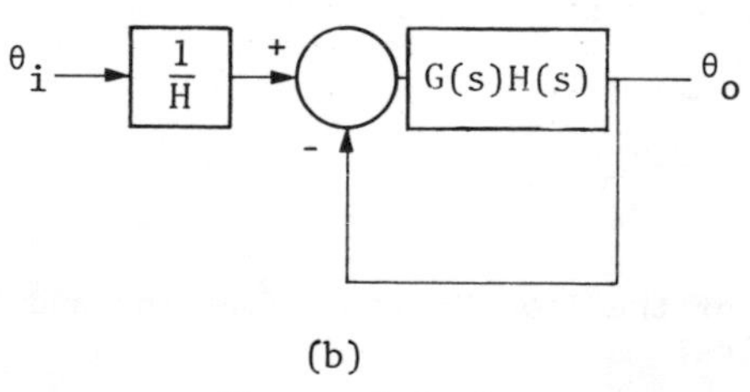

(b)

Figure 5.1

Consider the system of fig. 5.1. The closed-loop transfer function is

$$\frac{\theta_o}{\theta_i}(s) = \frac{G(s)}{1 + G(s)H(s)} \tag{5.1}$$

The sinusoidal response may be obtained by replacing the Laplace operator (s) by (jω)

$$\frac{\theta_o}{\theta_i}(j\omega) = \frac{G(j\omega)}{1 + G(j\omega)H(j\omega)} \tag{5.2}$$

For stability, all the roots of the characteristic equation

$$1 + G(s)H(s) = 0 \tag{5.3}$$

must lie in the left-hand half of the s plane. The Nyquist stability criterion is one that relates the open-loop frequency response $G(j\omega)$ $H(j\omega)$ to the number of poles and zeros of $1 + G(s)H(s)$ that lie in the right-hand half of the s plane. For a closed-loop system to be stable it is necessary and sufficient that the contour of the open-loop frequency response $G(j\omega)H(j\omega)$ plotted as a polar diagram describe a number of counterclockwise encirclements of the point (-1, j0) as ω varies from $-\infty$ to $+\infty$ not less than the number of poles of $G(s)H(s)$ with positive real parts.

A Nyquist plot may be chosen as an analytical method to obtain information about system stability in preference to using Routh's criterion. This criterion is explained in examples 5.3 and 5.5 and is often inadequate because normally it can only be used to determine 'absolute' stability since it is applied to a system whose characteristic equation is a finite polynomial in s. No such restriction applies to Nyquist's criterion, which yields exact results about both absolute and relative stability.

The term 'relative stability' is used to indicate the degree of stability of a system and is associated with the nearness of the open-loop frequency plot to the (-1, j0) point. Two quantitative measurements of gain margin and phase margin can then be made to determine this degree of stability; these are fully explained in example 5.1.

Example 5.1

The following experimental results were obtained from an open-loop frequency response test of an automatic control system.

ω rad s^{-1}	4	5	6	8	10
Gain	0.66	0.48	0.36	0.23	0.15
Phase angle(°)	-134	-143	-152	-167	-180

Plot the locus of the loop transfer function and measure the gain and phase margins.

By what factor should the gain be increased so that the maximum closed-loop gain is 1.4 and what would then be the gain and phase margins?

The given information is shown in fig. 5.2. By definition, *gain margin* is the amount by which the open-loop gain must be increased, at the frequency at which the phase angle is -180°, in order to enclose the -1 point. In this case 1/0.15 = 6.7.

The *phase margin* is the additional open-loop phase shift required at the frequency where the gain is unity, in order that the -1 point shall be enclosed. In this case, by drawing the unity circle as shown in fig. 5.2, the phase margin is measured as 59°.

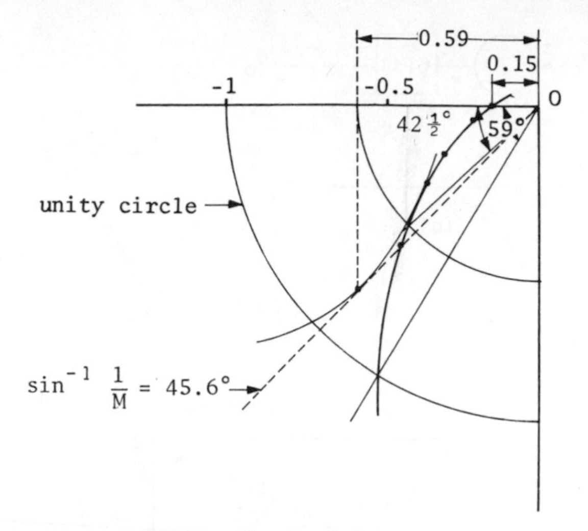

Figure 5.2

For the second part of the problem, a line is drawn from the origin at angle $\sin^{-1}(1/M)$ and a circle is found with centre on the real axis and tangential to both the plot and this line. A perpendicular to the real axis from the point of contact of the circle with the line crosses the real axis at -0.59, so that the gain should be increased by 1/0.59 or 1.75. The radius of the unity circle becomes 0.59 in order to measure the phase margin. From fig. 5.2 the phase margin is $42\frac{1}{2}°$, and the gain margin is 6.7/1.75 or 3.8.

Example 5.2

Explain how the stability of a servo system may be examined using a Nyquist plot. Hence derive expressions for the radius and position of the circles that represent the lines of constant closed-loop gain, in terms of the modulus M.

Consider the system, with unity feedback, shown in fig. 5.3a to have an open-loop frequency response as shown in fig. 5.3b. If the loop gain magnitude is less than unity when the phase angle is 180°, the system will be stable.

The closed-loop frequency response is given by

$$\frac{\theta_o}{\theta_i}(j\omega) = \frac{G(j\omega)}{1 + G(j\omega)} = |M| \underline{\lfloor \alpha}$$

The magnitude and phase angle characteristics of the closed-loop frequency response of a unity feedback control system can be determined directly from the polar plot of $G(j\omega)$ as shown in fig. 5.3b. This is accomplished by first drawing lines of constant magnitude, called M-circles and if necessary, lines of constant phase, called N-circles, directly on to the $G(j\omega)$ plane. The intersection of the polar plot with a particular M-circle yields the value of M at the frequency ω of $G(j\omega)$ at the point of intersection.

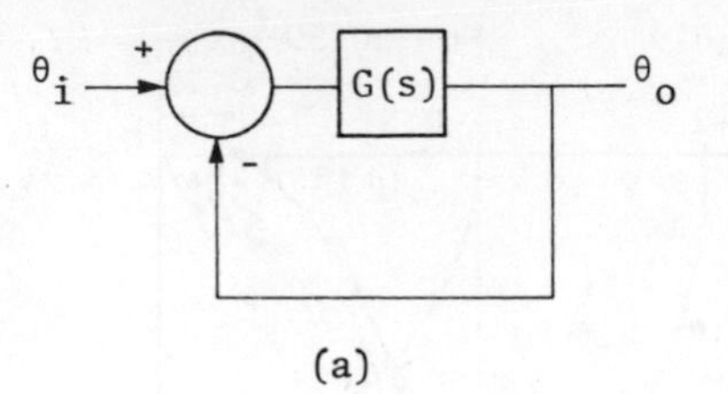

(a)

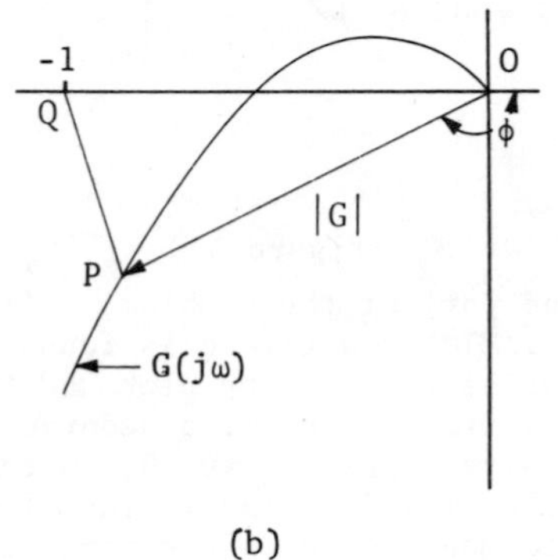

(b)

Figure 5.3

The maximum or peak value of M is given by the largest value of M_{max} of the M-circle(s) that is tangential to the polar plot.

The modulus is

$$M = \frac{|\theta_o|}{|\theta_i|} = \frac{|OP|}{|PQ|}$$

Let the point P have co-ordinates (x, y)

$$M = \frac{\sqrt{(x^2 + y^2)}}{\sqrt{(1 + x)^2 + y^2}}$$

therefore

$$M^2(1 + 2x + x^2 + y^2) = x^2 + y^2$$

$$x^2 + y^2 + \frac{M^2(2x + 1)}{M^2 - 1} = 0$$

or $$\left[x + \frac{M^2}{M^2 - 1}\right]^2 + y^2 = \frac{M^2}{(M^2 - 1)^2}$$

The locus of constant gain is a circle, of radius $M/(M^2 - 1)$ and with centre at $x = -M^2/(M^2 - 1)$, $y = 0$.

Fig. 5.4 shows a family of M-circles for $1 < M > 1$.

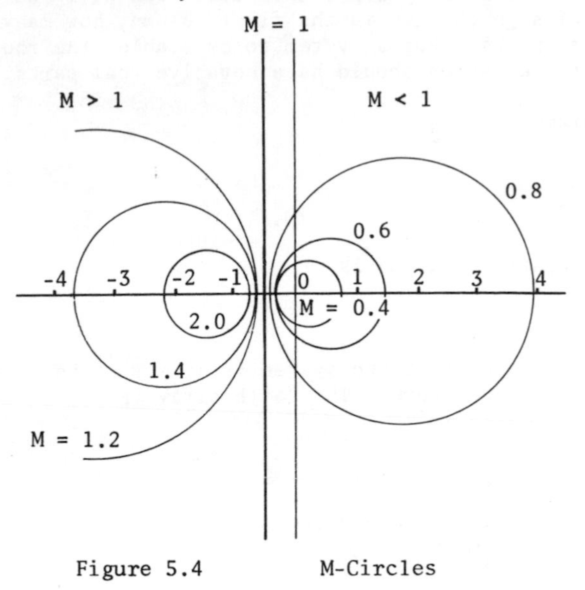

Figure 5.4 M-Circles

Example 5.3

Define the characteristic equation of a closed-loop system. Explain how it may be employed in assessing the stability of the system.

The open-loop transfer function of three systems are

$$\frac{K_1}{s^2(1 + 0.1s)}$$

$$\frac{K_2}{(1 + s)(1 + 0.1s)}$$

$$\frac{K_3 + s}{s^2(1 + 0.1s)}$$

Determine the range of values (if any) of K for which each system is stable when closed through unity negative feedback.
(C.E.I. E.F.C., 1973.)

For any system with forward transfer function G(s) and feedback transfer function H(s) the characteristic equation is obtained from the closed-loop transfer function

$$\frac{\theta_o}{\theta_i}(s) = \frac{G(s)}{1 + G(s)H(s)}$$

i.e. $1 + G(s)H(s) = 0$

is the equation required. The roots of this equation are examined by use of the Routh array since this criterion will indicate, from the number of sign changes in the first column, how many roots have positive real parts. For a system to be stable, the roots of the characteristic equation should have negative real parts.

For the examples

$$\frac{K_1}{s^2(1 + 0.1s)}$$

The characteristic equation is

$$0.1s^3 + s^2 + K_1 = 0$$

Inspection indicates that the system is likely to be unstable since the equation has no s term. The Routh array is

$$\begin{array}{lll} s^3 & 0.1 & 0 \\ s^2 & 1 & K_1 \\ s^1 & 0 - 0.1K_1 & \\ s^0 & K_1 & \end{array}$$

Two changes of sign, so the system is unstable for all K_1 values.

$$\frac{K_2}{(1 + s)(1 + 0.1s)}$$

The characteristic equation is

$$0.1s^2 + 1.1s + K_2 + 1 = 0$$

The Routh array is

$$\begin{array}{lll} s^2 & 0.1 & K_2 + 1 \\ s^1 & 1.1 & \\ s^0 & \dfrac{1.1(K_2 + 1) - 0}{1.1} & \end{array}$$

For positive values of K_2 the system is always stable.

$$\frac{K_3 + s}{s^2(1 + 0.1s)}$$

The characteristic equation is

$$0.1s^3 + s^2 + s + K_3 = 0$$

The Routh array is

s^3	0.1	1
s^2	1	K_3
s^1	$1 - 0.1K_3$	
s^0	K_3	

Therefore $K_3 < 10$ for the system to be stable.

Example 5.4

The forward path transfer function of a direct feedback system is given by

$$G(s) = \frac{K}{s(1 + sT_1)(1 + sT_2)}$$

Determine an expression for the maximum value of K to ensure stability.

For this problem Routh's criterion can be applied to the characteristic equation of the system to determine the polarity of the roots and so yield the limiting value of K, from the Routh array.

The characteristic equation is

$$s^3T_1T_2 + s^2(T_1 + T_2) + s + K = 0$$

The array is

s^3	T_1T_2	1
s^2	$T_1 + T_2$	K
s^1	$\frac{(T_1 + T_2) - KT_1T_2}{T_1 + T_2}$	
s^0	K	

The maximum value of K is $(T_1 + T_2)/T_1T_2$.

Example 5.5

A control system has an open-loop frequency response $G(j\omega)$ with properties as follows.

ω rad s^{-1}	2	3	4	5	6
$\lvert G(j\omega)\rvert$	2.7	1.7	0.97	0.63	0.4
$\angle G(j\omega)°$	-115	-126	-138	-150	-163

Plot the open-loop frequency response.

For the closed-loop system, find (a) the maximum value of the magnification M for a sinusoidal input, (b) the damped natural frequency for a step input, (c) the bandwidth of the system.

The frequency response is shown in fig. 5.5 and is tangential to the $M = 1.4$ circle at a frequency $\omega = 4$ rad s^{-1}.

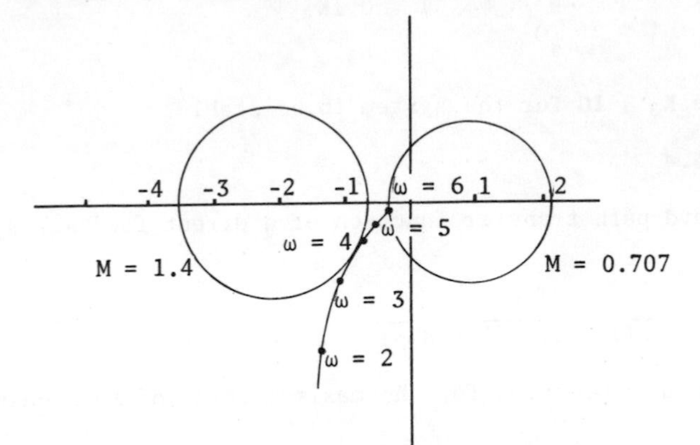

Figure 5.5

For a second-order system, it can be shown that

$$M_{max} = \frac{1}{2\zeta\sqrt{(1 - \zeta^2)}}$$

$$\omega_{max} = \omega_n\sqrt{(1 - 2\zeta^2)}$$

Therefore by substitution, $\zeta = 0.39$ and $\omega_n = 4.8$ rad s^{-1} so that

$$\omega_D = 4.8\sqrt{(1 - 0.39^2)} = 4.43 \text{ rad s}^{-1}$$

The bandwidth of a system is defined as the frequency at which the magnitude of the closed-loop frequency response is 0.707 of its magnitude at $\omega = 0$. Thus by drawing the $M = 0.707$ circle in fig. 5.5, which intersects the plot at $\omega = 6$ rad s^{-1}, this can be taken as the bandwidth if it is assumed that the closed-loop gain is unity at $\omega = 0$.

Example 5.6

The output/error transfer function of a system with a unity feedback loop is

$$G(s) = \frac{k_1 k_2}{s^2(1 + sT)}$$

Sketch the Nyquist locus and thence deduce that the system is unstable on closed-loop.

The system is now stabilised by adding a feedback $(-\lambda)$ across a section $k_2/s(1 + sT)$ of $G(s)$, with λ such that the gain margin is

0.5. Obtain the modified output/error transfer function of the system and show that for the specified condition, (a) $\lambda = 2Tk_1$ and (b) the value of |error/input| at a low frequency ω is $2T$.
(C.E.I. E.F.C., 1970.)

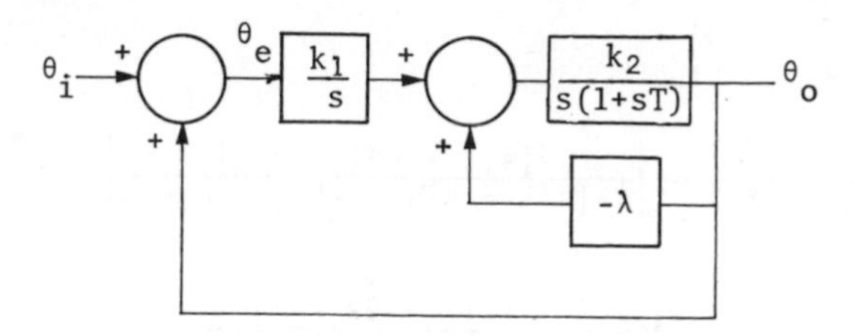

Figure 5.6

The expression for G(s) has a type-2 classification indicated by the (s^2) term in the denominator, and shows that at low frequencies the Nyquist locus is asymptotic to the negative real axis.

$$G(j\omega) = \frac{k_1k_2(-1 + j\omega T)}{\omega^2 + \omega^4T^2}$$

and the locus is shown in fig. 5.7 encircling the (-1, j0) point, which is the Nyquist criterion for instability.

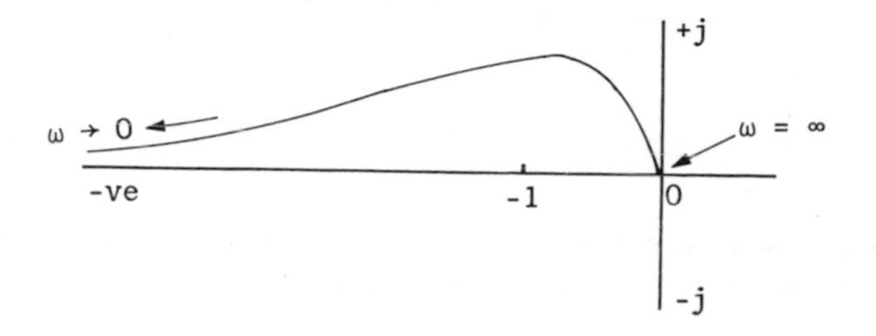

Figure 5.7

Introducing ($-\lambda$) feedback as shown in fig. 5.6

$$\frac{\theta_o}{\theta_e} = \frac{k_1}{s\left[\frac{s(1 + sT)}{k_2} + \lambda\right]} = G'(s)$$

For a gain margin of 0.5, the value of G'(s) is $0.5\angle{-180°}$, therefore

$$\frac{k_1k_2}{s[s(1 + sT) + \lambda k_2)]} = -0.5$$

In frequency mode

$$k_1k_2 = 0.5\omega^2(1 + j\omega T) - j\omega 0.5\lambda k_2$$

therefore

$k_1k_2 = 0.5\omega^2$ or $\omega^2 = 2k_1k_2$

and $0.5\omega^3T = \omega 0.5\lambda k_2$

(a) Therefore

$$\lambda = 2k_1T$$

$$\left|\frac{\text{Error}}{\text{Input}}\right| = \left|\frac{\theta_e}{\theta_i}\right| = \left|\frac{s[s(1 + sT) + \lambda k_2]}{\{s[s(1 + sT) + \lambda k_2] - k_1k_2\}}\right|$$

$$\frac{\theta_e}{\theta_i}(j\omega) = \frac{j\omega[j\omega(1 + j\omega T) + 2k_1k_2T]}{\{j\omega[j\omega(1 + j\omega T) + 2\,k_1k_2T] - k_1k_2\}}$$

$\omega \to 0$

$$\frac{\theta_e}{\theta_i} \simeq \frac{2T0.5\omega^2}{2T0.5\omega^2 + j\frac{k_1k_2}{\omega}} = \frac{\omega^2T}{T\omega^2 + j\frac{1}{2}\omega} = \frac{\omega T}{\omega T + j\frac{1}{2}}$$

$$\simeq 2T\omega\,\underline{|-90^\circ}$$

therefore (b)

$$\left|\frac{\theta_e}{\theta_i}\right| = 2T\omega$$

Problems

1. The open-loop transfer function of a control system depends on the angular frequency ω rad s^{-1} according to

$$G(j\omega) = \frac{K(9 - \omega^2 + j2\omega)}{3 - \omega^2 + j\omega(4 - \omega^2)}$$

where K is a real positive number. Determine the frequencies for which the transfer function is real. Sketch the locus of the function and hence find the values of K for which the system is (a) stable, and (b) unstable.

[0 rad s^{-1}, $\sqrt{5}$ rad s^{-1}, $\sqrt{6}$ rad s^{-1}; (a) $\frac{1}{2} > K > 1$, (b) $\frac{1}{2} < K < 1$]

2. (a) Derive the Nyquist stability criterion.

(b) Sketch the Nyquist diagrams and determine the stability of systems with open-loop transfer functions

(i) $G(s)H(s) = \dfrac{1}{s(s + 1)}$

(ii) $G(s)H(s) = \dfrac{1}{s^5(1 + s)}$

(C.E.I. S & C Eng., 1973) [(i) stable, (ii) unstable]

3. Construct a Nyquist diagram for the function

$$G(s) = \frac{K}{s(1 + 0.5s)(6 + s)}$$

and determine the value of K that will give the closed-loop system a value of $M_{max} = 1.4$.

[10.4]

4. The open-loop transfer function of a servo system is given by

$$G(s)H(s) = \frac{8}{s(1 + s0.4)(1 + s0.2)}$$

Establish that such a system is unstable in closed-loop connection.

If the system is stabilised by decreasing the gain to give a phase margin of 15°, evaluate the new gain margin and the required gain setting. Explain the procedure you adopt in these evaluations.

[1.68, 4.45]

5. State the Routh criterion for the stability of a closed-loop system. Discuss the limitations of this technique.

Use the criterion to calculate the maximum value of A for which the control system shown in fig. 5.8 is stable.

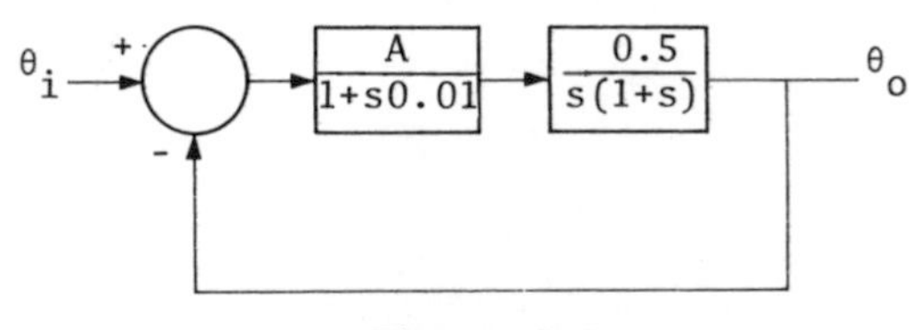

Figure 5.8

[A = 202]

6. A frequency test on the opened loop of a small system gave the following results

ω rad s^{-1}	2.5	3	3.5	4	4.5
Real part $G(j\omega)$	-2.0	-1.75	-1.5	-1.25	-1.0
Imaginary part $G(j\omega)$	-2.5	-1.67	-1.17	-0.8	-0.5

Plot the Nyquist diagram and show that M_{max} has a value of 2.3 when $\omega = 4.7$. Determine the change in the system gain required to give an M_{max} value of 1.5. State the new resonant frequency.

[0.5, 3.5 rad s^{-1}]

7. State the Nyquist stability criterion. Sketch the Nyquist diagram for a system whose open-loop transfer function is

$$G(s) = \frac{1}{s(s + 2)(s - 1)}$$

Use the Nyquist stability criterion to deduce whether there is any value of feedback gain k (positive or negative) that will stabilise the system shown in fig. 5.9.

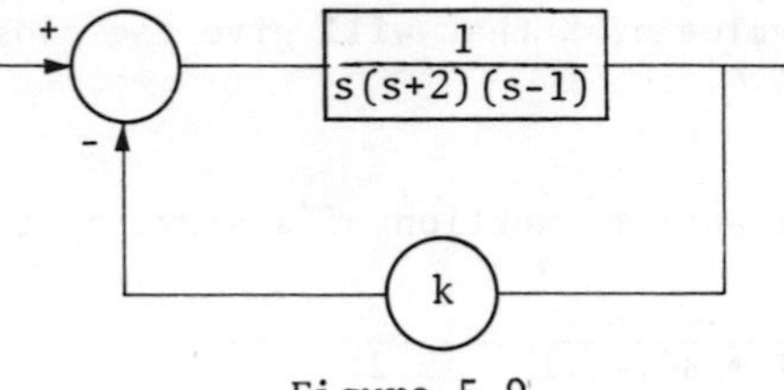

Figure 5.9

(C.E.I. C.S.E., 1975) [k > 2]

6 THE BODE DIAGRAM

The frequency-response characteristic as shown on a Nyquist diagram is easy to visualise and provides a convenient method for system evaluation. However, the actual plotting is tedious and where the systems require analytical modifications and performance checks, the method becomes somewhat impractical. The problem is simplified by the preparation of logarithmic diagrams.

The logarithms of the modulus $|GH|$ and the argument $\underline{|GH}(j\omega)$ are plotted separately against the logarithm of frequency using Cartesian axes.

The advantages are as follows.

(1) Multiplication of vectors is replaced by the addition of logarithmic curves, so if

$$G(j\omega) = \frac{K}{j\omega(1 + j\omega T)}$$

then $20 \log_{10}|G| \text{dB} = 20 \log_{10} K - 20 \log_{10}|\omega| - 20 \log_{10}|1 + j\omega T|$

Multiplying the gain factor K by a factor α simply moves the whole diagram through $\log \alpha$.

(2) The loci for $(1 + j\omega T)$ and $1/(1 + j\omega T)$ can be represented by straight-line asymptotes. Thus the expression

$$\left|\frac{1}{1 + j\omega T}\right| = \frac{1}{(1 + \omega^2T^2)^{\frac{1}{2}}} \to 1 \text{ as } \omega \to 0$$

and tends to

$$\frac{1}{\omega T} \text{ as } \omega \to \infty$$

Hence

$$20 \log_{10}\left|\frac{1}{1 + j\omega T}\right| \to 0 \text{ as } \omega \to 0$$

and tends to

$$20 \log_{10} \frac{1}{T\omega} \text{ as } \omega \to \infty$$

These straight lines meet where $\log(1/T) - \log \omega = 0$, i.e. at $\omega = 1/T$. This is called the 'corner frequency' or 'break frequency' and at that frequency the true value of $|1/(1 + j\omega T)| = 1/\sqrt{2}$ which in logarithmic terms is equal to 3 dB. This is the maximum error, for a single term of this type, when using straight-line asymptotes, while an error of 1 dB occurs at ± 1 octave to the corner frequency. The curve for the phase angle $\underline{|G(j\omega)}$ is obtained by adding the phase angles for the individual terms, i.e. $- \underline{|90°} - \tan^{-1} \omega T$.

Bode plots will clearly illustrate the relative stability of a system. In fact, gain and phase margins are often defined in terms of Bode plots, since they can be determined for a particular system with a minimum of effort, particularly when experimental frequency-response data are available.

Example 6.1

Prepare a Bode diagram for the open-loop transfer function

$$G(j\omega) = \frac{5}{j\omega(1 + j\omega 0.6)(1 + j\omega 0.1)}$$

and determine the phase margin and gain margin.

The calculation for the gain in decibels is carried out as follows

$$20 \log|G| = 20 \log_{10}5 - 20 \log_{10}\omega - 20 \log_{10}|(1 + j\omega 0.6)| - 20 \log_{10}|(1 + j\omega 0.1)|$$

$$20 \log_{10}5 = 20 \times 0.698 = 14 \text{ dB}$$

There are two corner frequencies at 1/0.6 or 1.67 rad s^{-1} and 1/0.1 or 10 rad s^{-1}. The gain curve is shown in fig. 6.1, with the plots of each factor added together.

Similarly the argument is

$$\underline{|G(j\omega)} = - \underline{|90°} - \underline{|\tan^{-1}} \omega 0.6 - \underline{|\tan^{-1}} \omega 0.1$$

Therefore the following table can be prepared

ω	1	2	3	4	5
$\underline{|G(j\omega)}°$	-127	-152	-168	-179	-188

The phase-angle curve is also plotted on fig. 6.1, showing that the gain margin is 6 dB and the phase margin 16°, i.e. the system is stable in closed loop but has a very poor transient performance.

Note For a phase margin of 30°, $M_{max} \simeq 2$ and this for a second-order system would correspond to a damping ratio of approximately 0.25, which in turn, corresponds, for a unit step input, to an overshoot of 40 per cent.

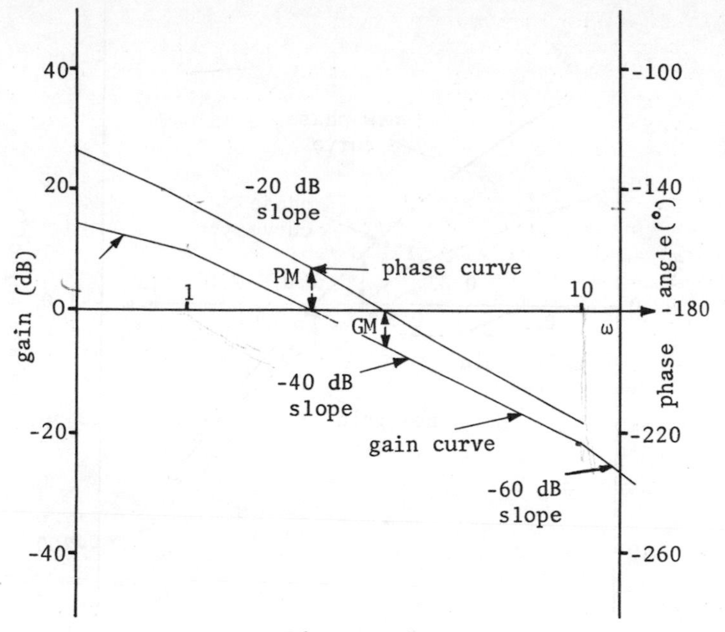

Figure 6.1

Example 6.2

The open-loop transfer function of a control system is

$$G(s)H(s) = \frac{1}{s(1 + 0.5s)(1 + 2s)}$$

(a) Determine approximate values of the gain and phase margins.

(b) If a phase-lag element with transfer function $(1 + 3s)/(1 + 5s)$ is inserted in the forward path, by about how much must the gain be changed to keep the gain margin unchanged?

(C.E.I. S & C Eng., 1973)

The transfer function of the system in frequency form is

$$G(j\omega)H(j\omega) = \frac{1}{j\omega(1 + j\omega 0.5)(1 + j\omega 2)}$$

and has two corner frequencies at 2 and 0.5 rad s^{-1}. The gain curve is shown in fig. 6.2. From this, the important frequencies can be chosen in order to calculate the angles for the phase curve.

ω rad s^{-1}	0.2	0.5	1	2
$\underline{/\phi°}$	-118	-149	-180	-211

When this has been plotted as shown in fig. 6.2, the gain and phase margins are measured as 6 dB and 17° respectively.

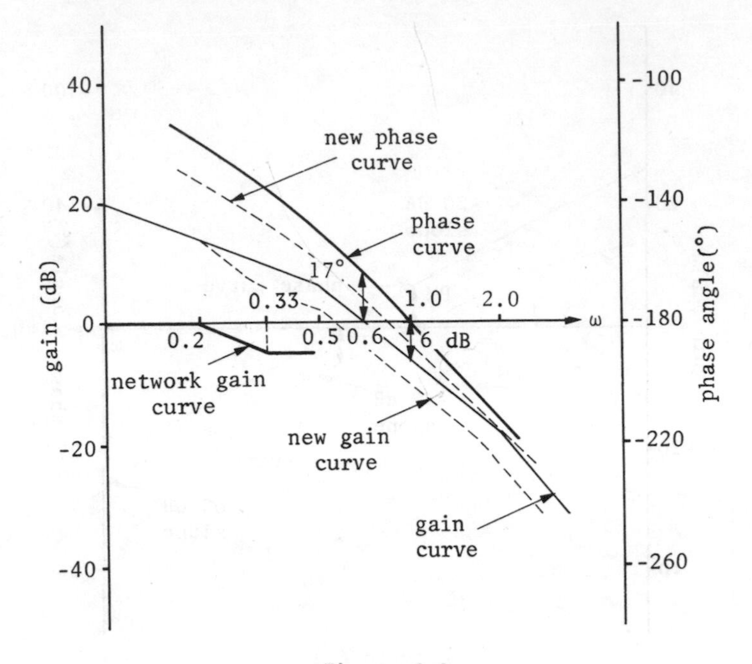

Figure 6.2

Now the gain and phases for the network are added to the diagram: there are two corner frequencies at 0.33 and 0.2 rad s^{-1} respectively and the gain curve is also shown on fig. 6.2.

Next evaluate the phase angles for the network for the same values of ω as before

ω rad s^{-1}	0.2	0.5	1	2
phase angle network °	-14	$-11\frac{1}{2}$	$-6\frac{1}{2}$	-4

Therefore by adding both gain curves and both phase-angle curves together, the new system response is obtained.

To keep the gain margin unchanged at 6 dB the gain must be changed by 1.5 dB.

Example 6.3

A system has an open-loop transfer function $k/[s^2(1 + sT_2)]$ relating the output to the input variables. The system is stabilised by a phase-advancing network having a transfer function $k[1 + sT_1]/[1 + sT_2]$. The value of T_1 is such that, if T_2 is ignored, the closed-loop response is critically damped. Sketch a straight-line approximation to the open-loop gain/frequency characteristic, and thus show that the condition for the system to be stable is $T_1 > 4T_2$.

Determine the approximate value of the frequency at which the gain ratio is unity.

The following data relate the phase angle ϕ, of a simple transfer lag to the normalised frequency ωT.

ωT	0.1	0.2	0.5	1.0	2.0	5.0	10.0
$\phi°$	-6	-11	$-26\frac{1}{2}$	-45	$-63\frac{1}{2}$	-79	-84

By using this information, or otherwise, determine the phase margin when $T_1 = 10T_2$.

(C.E.I. E.F.C., 1968)

The transfer function of the system including the network is

$$G(j\omega) = \frac{k^2(1 + sT_1)}{s^2(1 + sT_2)^2}$$

The closed-loop transfer function is

$$\frac{\theta_o}{\theta_i} = \frac{k^2(1 + sT_1)}{s^2(1 + sT_2)^2 + k^2(1 + sT_1)}$$

With $T_2 = 0$ the characteristic equation is

$$s^2 + sT_1k^2 + k^2 = 0$$

For critical damping $k^4T_1^2 = 4k^2$, or $k = 2/T_1$. On substituting this value of k, the open-loop transfer function is

$$G(s) = \frac{4(1 + sT_1)}{s^2T_1^2(1 + sT_2)^2}$$

and the characteristic equation is now

$$s^4T_1^2T_2^2 + s^3 2T_1^2T_2 + s^2T_1^2 + s4T_1 + 4 = 0$$

The Routh array for this equation is

$$\begin{array}{lll}
T_1^2T_2^2 & T_1^2 & 4 \\
2T_1^2T_2 & 4T_1 & \\
T_1^2 - 2T_1T_2 & 4 & \\
\dfrac{4T_1^2 - 16T_1T_2}{T_1 - 2T_2} & & \\
4 & &
\end{array}$$

From the fourth figure in the first column of the array, $T_1 = 4T_2$ for critical stability.

To draw the Bode diagram gain curve, let $\omega T_2 = 1$ and if $T_2 = 1$, then $T_1 = 4$ and the second corner frequency $\omega = 1/T_1 = 0.25$. The

Bode diagram is shown in fig. 6.3, from which the frequency for a gain ratio of unity is seen to be $\omega = 1/T_2$.

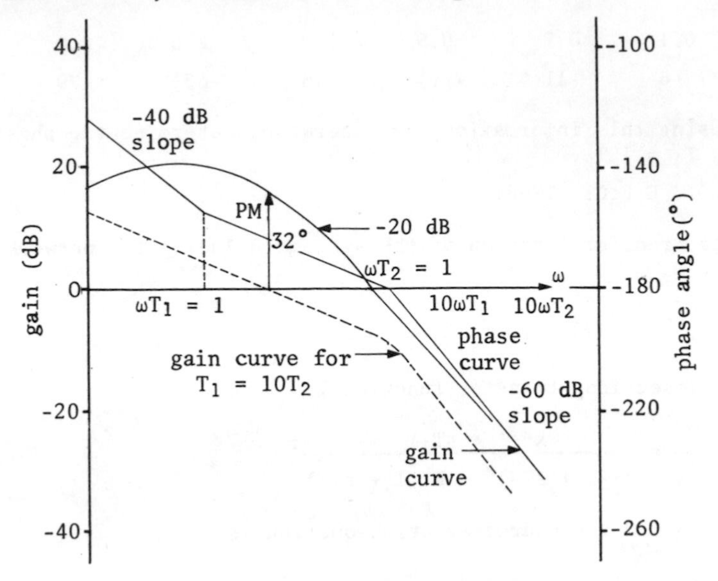

Figure 6.3

For the last part of the question, when $T_1 = 10T_2$, let $\omega T_2 = 1$ as before and $T_2 = 1$ so that $T_1 = 10$ and the second corner frequency $\omega = 1/T_1 = 0.1$; thus a new gain curve can be drawn, and using the the given phase-angle information, the corresponding phase curve is also shown in fig. 6.3, i.e.

$$G(j\omega) = \frac{4(1 + j\omega T_1)}{(j\omega T_1)(1 + j\omega T_2)^2}$$

with $T_1 = 10T_2$. The phase margin is 32°

Example 6.4

The Bode diagram for a unity feedback control system is given in fig. 6.4. Deduce the forward transfer function for the system and the approximate form of the phase-frequency curve. Hence comment on the stability, transient and steady-state performance of the system when operating in the closed-loop mode.

The given gain curve is reproduced in fig. 6.5 together with the phase curve after deducing the expression for the forward transfer function as

$$\frac{K(1 + j\omega 0.5)(1 + j\omega 0.25)}{j\omega(1 + j\omega 0.125)(1 + j\omega 0.042)(1 + j\omega 0.028)}$$

from fig. 6.5, deduce that $20 \log_{10} K = 18$, i.e. $K = 7.94$. Thus the phase margin is 50° and the gain margin is 24 dB.

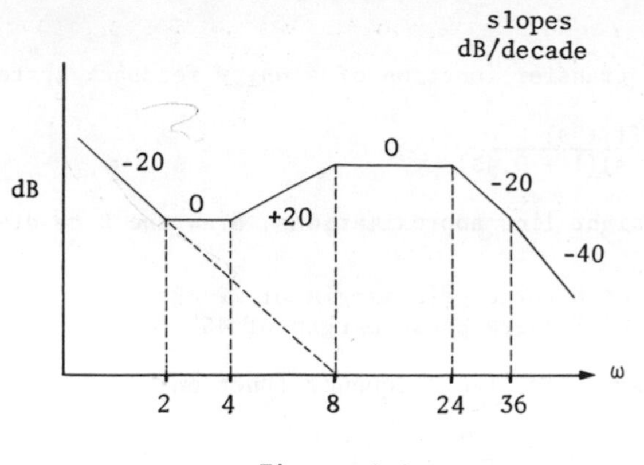

Figure 6.4

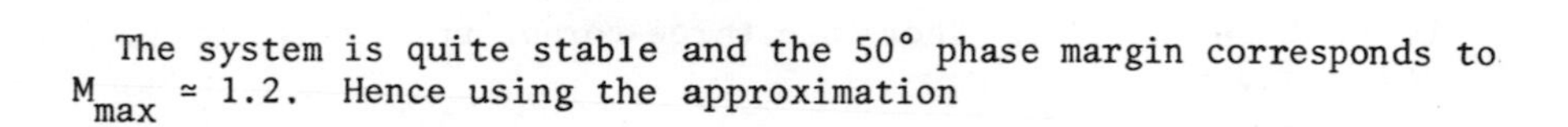
The system is quite stable and the 50° phase margin corresponds to $M_{max} \simeq 1.2$. Hence using the approximation

$$M_{max} = \frac{1}{2\zeta(1 - \zeta^2)^{\frac{1}{2}}}$$

$$\zeta = 0.47$$

so that the transient is oscillatory with an overshoot of approximately 19 per cent in response to a unit step input.

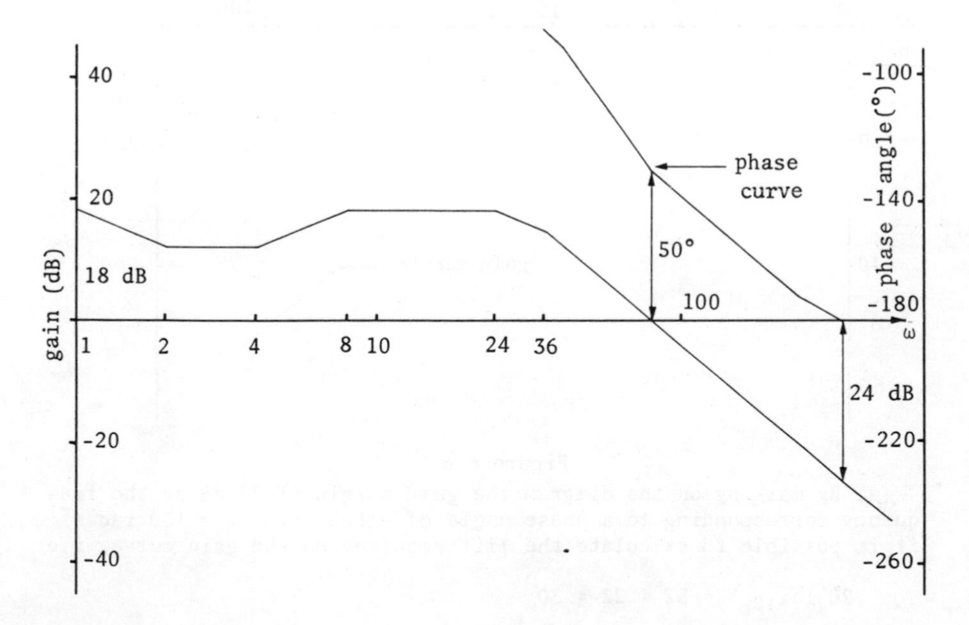

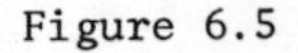
Figure 6.5

Example 6.5

The open-loop transfer function of a unity feedback system is

$$\frac{K(1 + s)}{s(1 + 0.1s)(1 + 0.4s)}$$

Using the straight-line approximations, draw the Bode diagrams and hence find

(a) the value of K for a gain margin of 22 dB
(b) the value of K for a phase margin of 45°

The Bode diagram for the frequency function

$$\frac{K(1 + j\omega)}{j\omega(1 + j\omega 0.1)(1 + j\omega 0.4)}$$

is shown in fig. 6.6. There are three corner frequencies at 1, 2.5 and 10 rad s^{-1}.

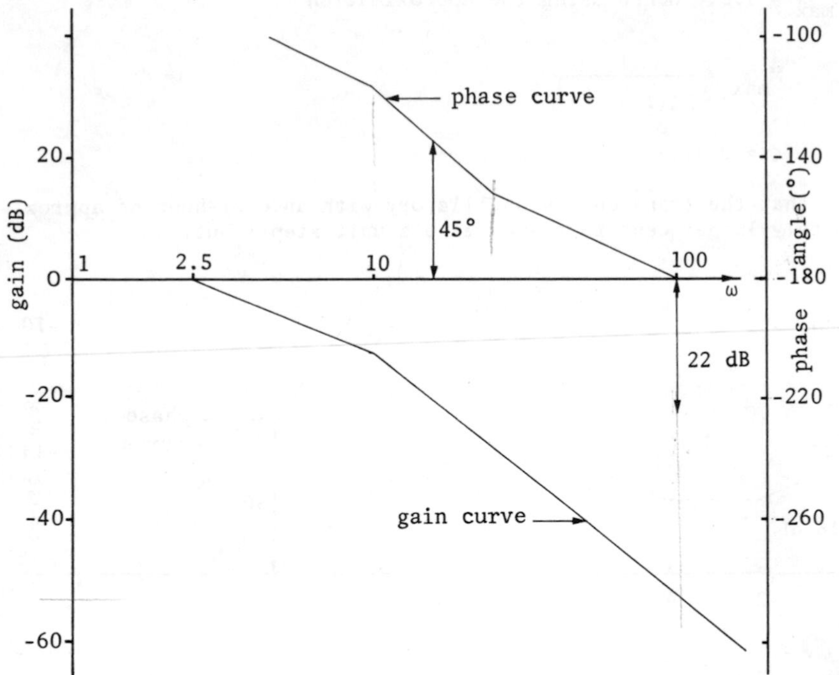

Figure 6.6

(a) By marking on the diagram the gain margin of 22 dB at the frequency corresponding to a phase angle of -180°, i.e. $\omega = 100$ rad s^{-1}, it is possible to calculate the lift required on the gain curve, i.e.

$$20 \log_{10} K = 52 - 22 = 30$$

thus K = 31.6

(b) By marking on the diagram the phase margin of 45° at the frequency of 16 rad s^{-1}, it is possible to read off the gain setting necessary to lift the gain curve to this level, i.e.

$$20 \log_{10} K = 20$$

thus K = 10

Example 6.6

A control system has an open-loop transfer function given by

$$G(j\omega) = \frac{K}{j\omega(1 + j\omega0.1)(1 + j\omega0.2)(1 + j\omega)}$$

with the following information given on phase angles

ω rad s^{-1}	0.6	1	2	3	4	10
$\angle G(j\omega)°$	-131	-152	-187	-209	-227	-283

Using a Bode plot approach, find, giving full details

(a) the value of the steady-state error coefficient that makes the system marginally stable on closed loop, and
(b) the value of the steady-state error coefficient that results in a gain margin of 10 dB and the corresponding phase margin.

The Bode diagram has been drawn and shown in fig. 6.7 with K = 1. The projection of the -20 dB slope will yield the steady-state error coefficient for a type-1 system.

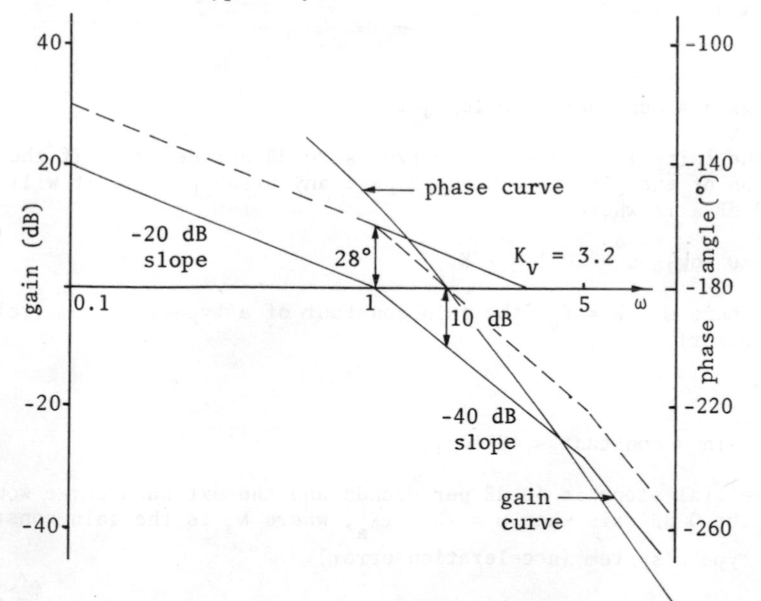

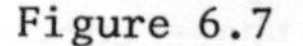
Figure 6.7

Readers are reminded that the initial slope of a Bode diagram yields information on the class of the system as well as the value of the loop gain - see fig. 6.8, where

$$G(j\omega) = \frac{K(1 + j\omega T_1)(1 + j\omega T_2) + \ldots}{(j\omega)^N(1 + j\omega T_A)(1 + j\omega T_B) + \ldots}$$

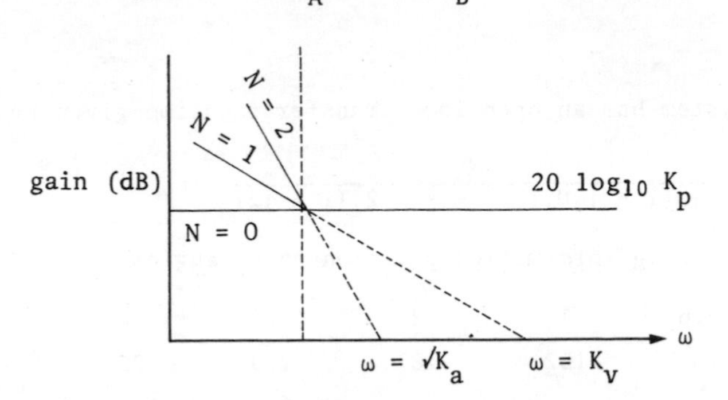

Figure 6.8

At frequencies less than the lowest corner value, the contribution of these linear factors will be 0/dB. Thus

$$\text{gain} = 20 \log_{10} K - 20\,N \log_{10} \omega$$

$N = 0$

$$\text{gain} = 20 \log_{10} K_p$$

$N = 1$

$$\text{gain} = \text{constant} - 20 \log_{10} \omega$$

and the initial slope of the curve is 20 dB per decade. If the first portion of the plot is continued past any break points, it will meet the 0 dB axis where

$$20 \log_{10} \omega = 20 \log_{10} K$$

i.e. where $\omega = K = K_v$, the gain constant of a type-1 system (velocity error).

$N = 2$

$$\text{gain} = \text{constant} - 40 \log_{10} \omega$$

The initial slope is 40 dB per decade and the extended curve would meet the 0 dB axis when $\omega = \sqrt{K} = \sqrt{K_a}$, where K_a is the gain constant of a type-2 system (acceleration error).

(a) For a marginal stable system, the gain curve in fig. 6.7 is lifted so that the gain and phase curves coincide at $\omega = 1.8$ rad s^{-1};

by projecting the -20 dB per decade slope until the 0 dB axis is cut, K_v is found to be 3.2.

(b) For a 10 dB gain margin, which is marked on fig. 6.7, it is evident that the steady-state error coefficient is 1 and the corresponding phase margin is approximately 28°.

Example 6.7

A control system with unity feedback has a closed-loop transfer function

$$\frac{\theta_o}{\theta_i}(s) = \frac{K}{0.1s^3 + 0.7s^2 + s + K}$$

Find the value of K that corresponds to the critical stability level of the system. If K = 4, plot the Bode diagram for the open-loop system and determine the phase and gain margins.

The characteristic equation is

$$0.1s^3 + 0.7s^2 + s + K = 0$$

Hence the array is

$$\begin{array}{lll} s^3 & 0.1 & 1 \\ s^2 & 0.7 & K \\ s^1 & \dfrac{0.7 - 0.1K}{0.7} & \\ s^0 & K & \end{array}$$

Hence from the s^1 term K = 7 for critical stability.

For a value of K = 4, the open-loop transfer function is

$$\frac{\theta_o}{\theta_e}(s) = \frac{4}{0.1s^3 + 0.7s^2 + s} = \frac{40}{s(s + 5)(s + 2)}$$

In terms of frequency

$$G(j\omega) = \frac{\theta_o}{\theta_e}(j\omega) = \frac{40}{j\omega(j\omega + 5)(j\omega + 2)} = \frac{4}{j\omega(1 + j\omega 0.2)(1 + j\omega 0.5)}$$

For the constant $20 \log_{10} 4 = 12$ dB. This expression has two corner frequencies at 5 and 2 rad s^{-1}, and the approximate gain curve is shown in fig. 6.9.

In plotting the phase angles, it is often simplest to calculate them for the whole function over the important range, say $\omega = 0.5$ to 6 rad s^{-1}.

ω	0.5	1.0	2.0	4.0	6.0
$\underline{\lvert G(j\omega)}$ °	-125	-128	-157	-192	-212

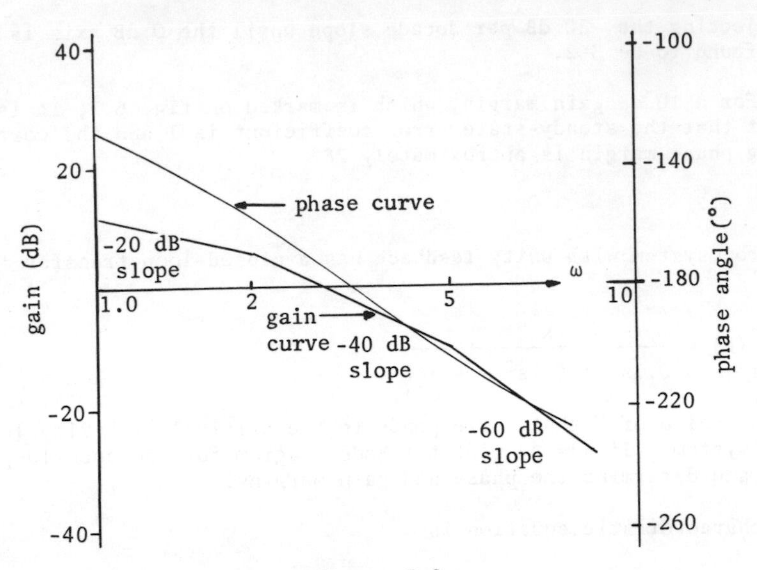

Figure 6.9

These results are plotted in fig. 6.9 giving a phase margin of 7° and a gain margin of 2.5 dB.

Example 6.8

The open-loop transfer function of a control system is given by

$$G(s) = \frac{7}{s(1 + 0.5s)(1 + 0.167s)}$$

Plot the Bode diagram and determine the gain and phase margins and the error constant.

A lag network having a transfer function $(1 + s\alpha T)/(1 + sT)$ $(\alpha < 1)$ is to be introduced as a series compensator to give a gain margin of at least 15 dB and a phase margin of 45°. Find suitable values of α and T.

$$G(j\omega) = \frac{7}{j\omega(1 + j\omega 0.5)(1 + j\omega 0.167)}$$

$20 \log_{10} 7 \equiv 16.9$ dB, and there are two corner frequencies of 2 and 6 rad s^{-1}. The Bode diagram for the basic system is shown in fig. 6.10. Again, in plotting the phase angles these have been calculated for the whole function over the important range, say $\omega = 0.1$ to 10 rad s^{-1}.

ω	0.1	1.0	5	10
$\angle G(j\omega)°$	-94	-126	-198	-228

The system is just unstable with a gain margin of -1 dB and a phase margin of -4°. The error coefficient is found by projecting the

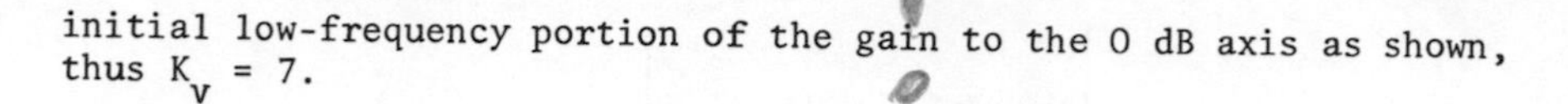

initial low-frequency portion of the gain to the 0 dB axis as shown, thus $K_v = 7$.

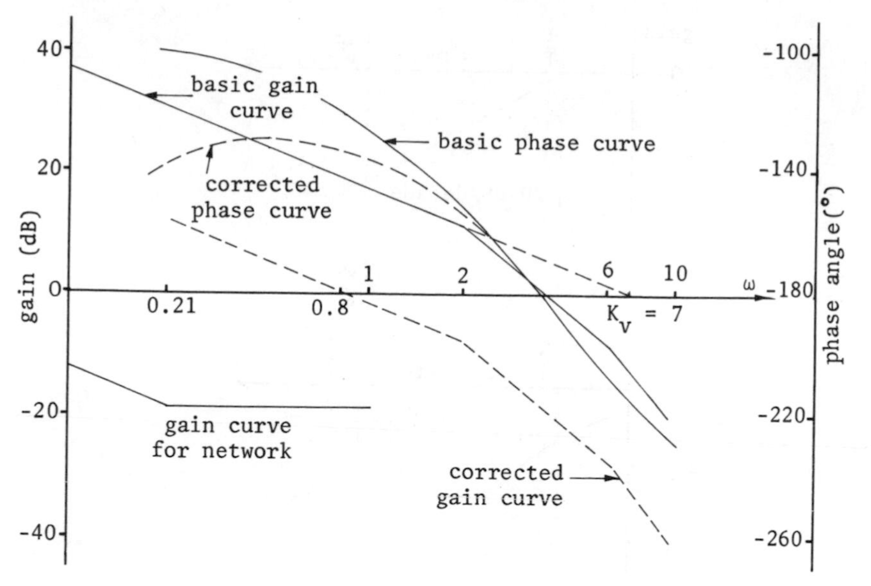

Figure 6.10

The asymptote of the added network has zero slope at low frequencies, changes to -20 dB per decade slope at $\omega_L = 1/T$ and back to zero slope at $\omega_H = 1/(\alpha T)$, as shown in fig. 6.11a. It introduces a phase lag of zero at high and low frequencies reaching $\tan^{-1}\alpha - 45°$ at the lower corner frequency ω_L, $45° - \tan^{-1}(1/\alpha)$ at the higher corner frequency ω_H and a maximum lag of $\tan^{-1}[\alpha - 1/(2\sqrt{\alpha})]$ at the geometric mean of the corner frequencies $\omega_{max} = 1/(T\sqrt{\alpha})$. The proof for these last two statements is left as an exercise.

The added network reduces the system gain by log α at all frequencies greater than $\omega = 1/(\alpha T)$.

This simple example in design means that the designer concentrates on moving the gain cross-over to a lower frequency and keeping the phase-shift curve almost unaltered at the cross-over frequency.

Suppose the new phase cross-over is assumed to be at $\omega = 3$ rad s^{-1}, then the gain reduction there will have to be 18.5 dB to give a gain margin of 15 dB. Hence $20 \log \alpha = -18.5$, giving $\alpha = 0.119$.

The corrected asymptote is now drawn as the dashed line in fig. 6.10. The gain cross-over is now seen to be at $\omega = 0.8$ rad s^{-1} and the additional phase lag at that frequency must not exceed 13° in order to give a phase margin of 45°, i.e. the phase angle of the network, at $\omega = 0.8$ with $\alpha = 0.119$, has to equal 13°.

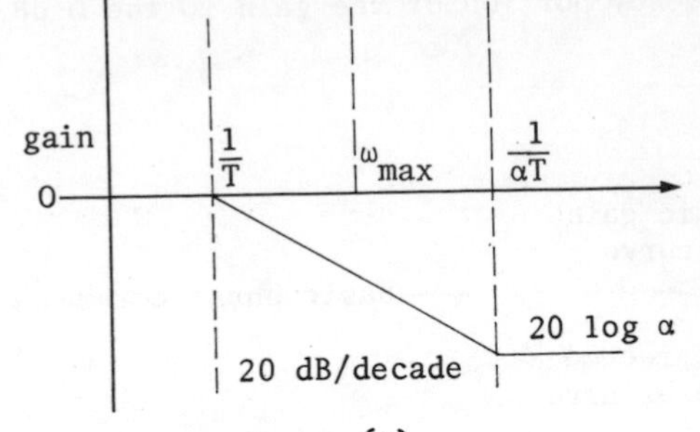

(a)

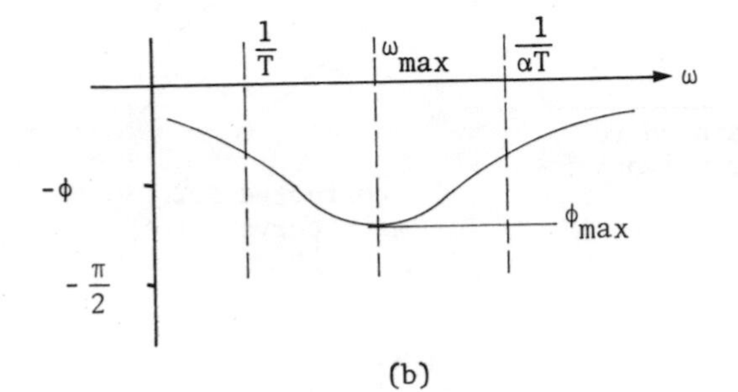

(b)

Figure 6.11

Let $v = \omega T$, $\theta_1 = \tan^{-1} \omega\alpha T$ and $\theta_2 = \tan^{-1} \omega T$. The phase lag $\theta = \theta_1 - \theta_2$ is given by

$$\tan \theta = \frac{\tan \theta_1 - \tan \theta_2}{1 + \tan \theta_1 \tan \theta_2} = \frac{v(\alpha - 1)}{1 + \alpha v^2}$$

$$\tan 13° = \frac{v(0.119 - 1)}{1 + 0.119v^2}$$

this leads to $v = 31.9$, therefore $T = 39.9$ s.

$$\omega_H = \frac{1}{\alpha T} = \frac{1}{0.119 \times 39.9} = 0.21 \text{ rad s}^{-1}$$

ω =	0.2	0.4	0.6	0.8	1.0
$\theta°$ =	-39	-24	-17	-13	-10

The information can be added to the Bode diagram, which shows the corrected gain and phase curves.

Problems

1. (a) Compare phase lag and phase lead compensation methods. Show how lags and leads are generated in practice.

(b) A phase lead network has a transfer function

$$G(s) = \frac{1 + 2s}{2(1 + s)}$$

Determine the maximum value of phase lead and the frequency at which it occurs. Sketch the Bode diagram for this network.

(C.E.I. S & C Eng., 1972) [19.5°, 0.707 rad s^{-1}]

2. Define the terms 'gain margin' and 'phase margin'. The asymptotic log modulus for the open path of a third-order system is shown in fig. 6.12. If the transfer function is of the form

$$KG(s) = \frac{K}{s(1 + sT_1)(1 + sT_2)}$$

determine the gain margin (in decibels).

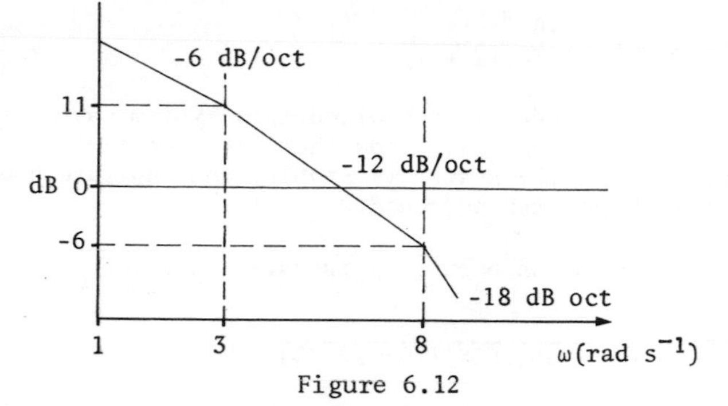

Figure 6.12

(C.E.I. S & C Eng., 1969) [-3 dB]

3. How can a frequency response be represented by a Bode diagram? Indicate what gain and phase margin are in this context.

Fig. 6.13 shows the asymptotic approximation to the variation of gain with frequency obtained from experimental results. Find the transfer function of the system, and sketch the phase-angle curve.

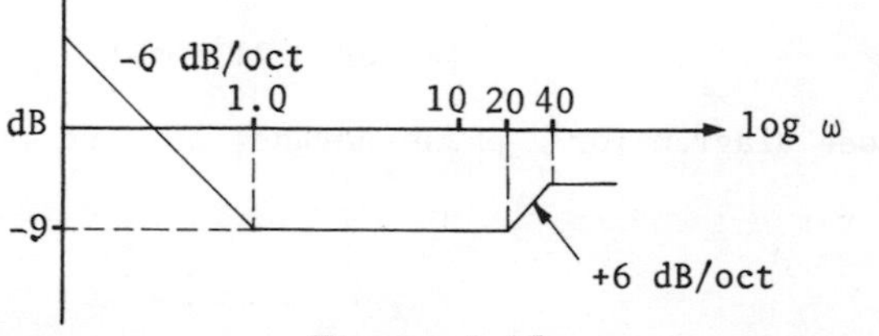

Figure 6.13

(C.E.I. S & C Eng., 1968)

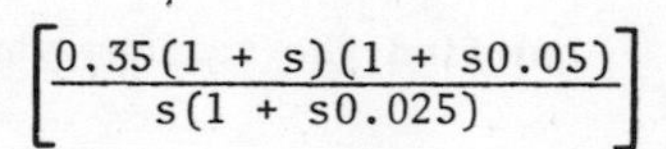

$$\left[\frac{0.35(1 + s)(1 + s0.05)}{s(1 + s0.025)}\right]$$

4. Give an account of the meaning of gain margin and phase margin in terms of (a) a Nyquist diagram, and (b) a Bode plot. Justify their determination from the Bode plot.

The open-loop transfer function of a unity feedback system is

$$\frac{K(1 + s)}{s(1 + 0.1s)(1 + 0.4s)}$$

Using the straight-line approximations, draw the Bode diagrams and hence find

(i) the value of K for a gain margin of 32 dB
(ii) the value of K for a phase margin of 60°

[10, 5]

5. Without detailed calculation of points, derive and sketch the main features of the Bode diagrams for the transfer function

$$G(s) = \frac{10}{s(0.2s + 1)(0.1s + 1)}$$

If unity negative feedback is applied to a system having the above forward transfer function G(s), show that it is possible to deduce from the diagram that the system is stable, and indicate how the phase and gain margins can be found.

6. A servo system has an open-loop transfer function

$$G(s)H(s) = \frac{40}{s(1 + s0.0625)(1 + s0.25)}$$

Plot the Bode diagram and determine the phase margin and gain margin and whether the servo is stable.

A phase-lag network is introduced into the servo system. Its trans fer function is

$$\frac{1 + s4}{1 + s80}$$

Prepare the Bode diagram for the modified system and find the new gain and phase margins.

[-7 dB, -21° unstable, 18 dB, 50°]

7. Plot the Bode diagram for a phase-advance circuit having a transfer function

$$\frac{1 + s0.2}{1 + s0.02}$$

and find the maximum phase-advance.

[55°]

8. The open-loop transfer function of a control system is given by

$$G(s) = \frac{20}{s(1 + s0.4)(1 + s0.2)}$$

With the aid of a Bode diagram show that the system can be stabilised by (a) decreasing the gain to give a phase margin of 20° (state the new gain setting and gain margin) and (b) using the gain-compensated phase-lead network of problem 7 for series compensation, phase margin and gain margin.

[(a) 6.0 dB, 8.0 dB; (b) 20° 14 dB]

7 NICHOLS CHART ANALYSIS

It was shown in chapter 5 that if a system had unity feedback, it was possible to obtain information on closed-loop response using M-circle data.

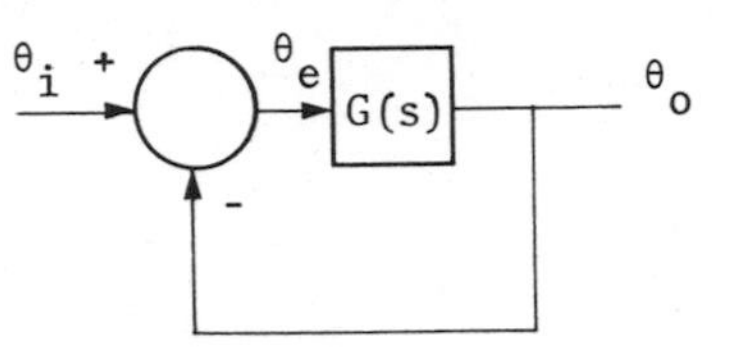

Figure 7.1

$$\frac{\theta_o}{\theta_i}(s) = \frac{G(s)}{1 + G(s)} \tag{7.1}$$

However, G(s) can be a high-order polynomial and the work involved in computing the closed-loop transfer function from the open-loop data can become tedious if attempted analytically. The Nichols chart provides a graphical method of system analysis that converts open-loop response to closed-loop response in one simple operation.

The chart comprises a set of rectangular co-ordinates of magnitude in dB against phase angle and against which open-loop data are plotted. Superimposed on the chart are a set of constant-gain and constant-phase contours (M- and N-contours) and against which closed-loop data can be read off. A typical chart is shown in fig. 7.2.

The contours and rectangular co-ordinates are related by eqn 7.1 and so the chart is only valid for unity feedback systems. In practice, this presents very little difficulty since fig. 5.1b shows how a transformation can be arranged, to provide a system with unity feedback without affecting the over-all performance. By using the 'new' open-loop transfer function G(s)H(s), the Nichols chart may be employed. However, the final closed-loop response obtained from the chart must be modified by the transfer function 1/H(s) in order to obtain the true response.

To use the chart, consider the open-loop frequency to have been determined so that it can be plotted against the rectangular co-ordinates. Gain in decibels is plotted in the vertical direction and phase angle in the horizontal direction. The closed-loop frequency response can then be read off at points of known frequency against the gain and phase contours. The relative stability features of gain and phase margins are also available from an inspection of the response curve.

From previous work, the gain margin is the open-loop gain when the phase angle is 180° and the phase margin is the difference between -180° and the actual phase angle when the open-loop gain is unity (i.e. 0 dB). The bandwidth can also be found at the frequency that corresponds to the intersection of the actual plotted curve with the -3 dB contour.

Furthermore, the general shape of the curve is, with some experience, a guide to the merits of the system design. The rapidity of phase and gain changes that occur at particular frequencies yields much insight into the performance, and can suggest appropriate corrective measures to improve the system response. In particular, the Nichols chart enables a quick method of synthesis since it is easy graphically to insert additional elements in series with the original open-loop transfer function. The transfer function of the new element and hence its frequency response will be known, so that its gain and phase quantities or characteristics can be added to the original response curve, thus giving the new closed-loop response.

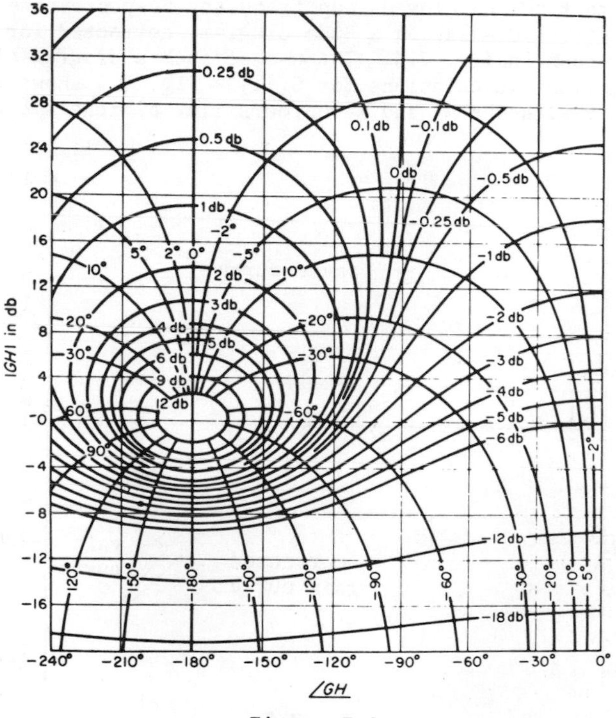

Figure 7.2

A simple change of open-loop gain merely requires the whole of the original open-loop response curve to be moved in the vertical direction by the appropriate number of decibels.

The Nichols chart method has two advantages over a Nyquist diagram. Firstly, a much wider range of magnitudes can be covered on one diagram since $|G(j\omega)|$ is plotted to a logarithmic scale. Secondly, the plot is obtained by the algebraic summation of the individual magnitude and phase angle contributions of its poles and zeros.

Although both these properties are applicable to a Bode diagram, the modulus and argument are included on a single chart rather than on two Bode diagrams.

Example 7.1

Consider the unity feedback system with an open-loop transfer function

$$G(s) = \frac{1}{s(s + 1)(0.5s + 1)}$$

Determine from a Nichols chart the value of M_{max}, ω_r and the bandwidth for the closed-loop system.

In order to find the closed-loop frequency response, the $G(j\omega)$ plot is constructed - say on a Bode diagram, corrected for the 3 dB error, and shown in fig. 7.3. The use of such a diagram eliminates lengthy numerical calculations for $G(j\omega)$. Fig. 7.4 shows the $G(j\omega)$ plot together with the M- and N-contours i.e. a Nichols chart.

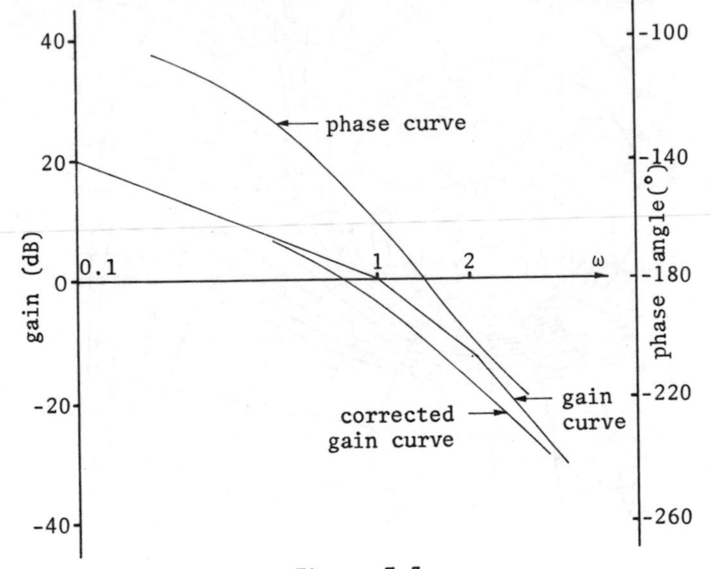

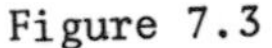

Figure 7.3

Since the $G(j\omega)$ plot is tangential to the M = 5 dB contour, the peak value of the closed-loop frequency response is M_{max} = 5 dB, while the corresponding resonant frequency is 0.8 rad s^{-1}. The bandwidth is given by the value of the frequency at the point where the $G(j\omega)$ plot and the M = - 3dB intersect, which is approximately 1.2 rad s^{-1}.

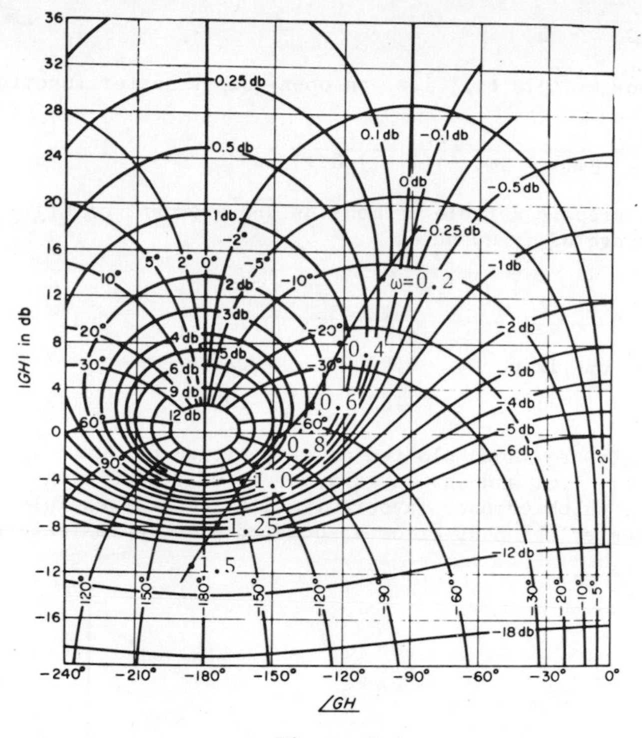

Figure 7.4

The gain margin is 10.5 dB while the phase margin is 32°.

If required, the closed-loop frequency response curves may be constructed from fig. 7.4 by reading the magnitude and phase angles at various frequency points, as shown in fig. 7.5.

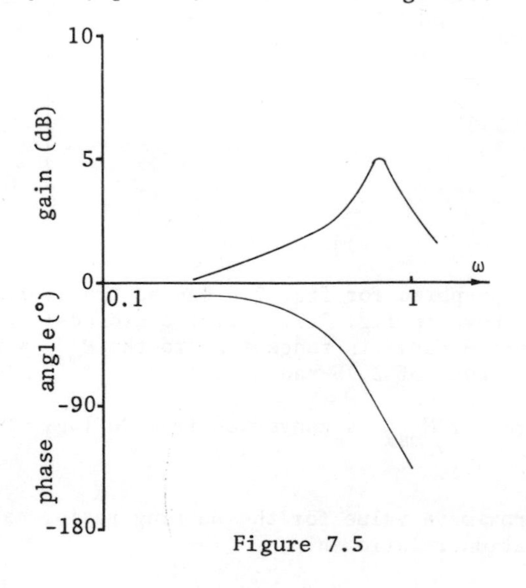

Figure 7.5

Example 7.2

Consider now example 6.1, i.e. an open-loop transfer function

$$G(j\omega) = \frac{5}{j\omega(1 + j\omega 0.6)(1 + j\omega 0.1)}$$

Firstly, prepare a table of modulus and argument of $G(j\omega)$ over a suitable range of frequencies

ω rad s^{-1}	1	2	3	4	5
$\lvert G \rvert$	4.27	1.57	0.775	0.447	0.282
$20 \log_{10} \lvert G \rvert$ dB	12.6	3.92	-2.22	-7	-11
$\underline{\lvert G(j\omega)}$ °	-127	-152	-168	-179	-188

These results are shown plotted on a magnitude dB scale against phase angle in fig. 7.6, and show a gain margin of 7.0 dB with a phase margin of 17°, which compare favourably with those of example 6.1 (6 dB, 16°) - remember the 3 dB error associated with approximate Bode lines.

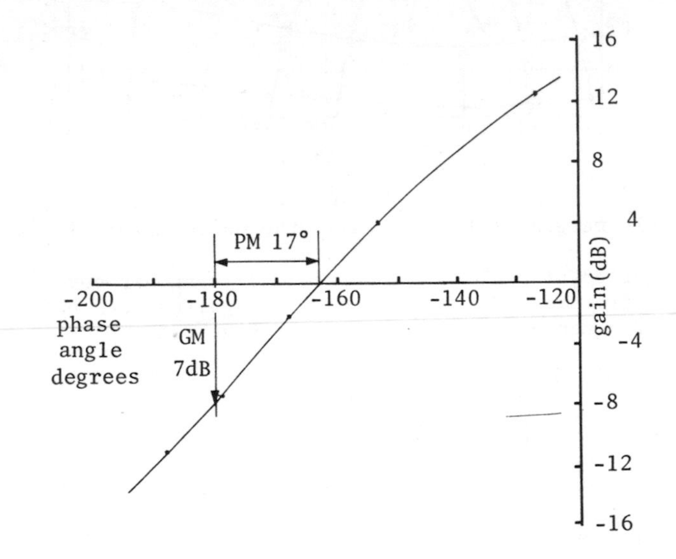

Figure 7.6

The information prepared for fig. 7.6 can be transferred to a Nichols chart as shown in fig. 7.7. For the closed-loop with unity feedback, the plotted curve is tangential to the M_{max} = 10 dB contour at a resonant frequency of 2.75 rad s^{-1}.

If the dB figure for M_{max} is converted from $20 \log_{10} M = 10$, then $M_{max} = 3.2$.

To find an approximate value for the damping ratio, make use of the second-order equation relationship

$$M_{max} = \frac{1}{2\zeta\sqrt{(1 - \zeta^2)}}$$

giving $\zeta = 0.16$ - a very underdamped system that would oscillate during the transient response to a step input for too long a period to be practical. The bandwidth is read off at the -3 dB contour as 4.2 rad s^{-1}.

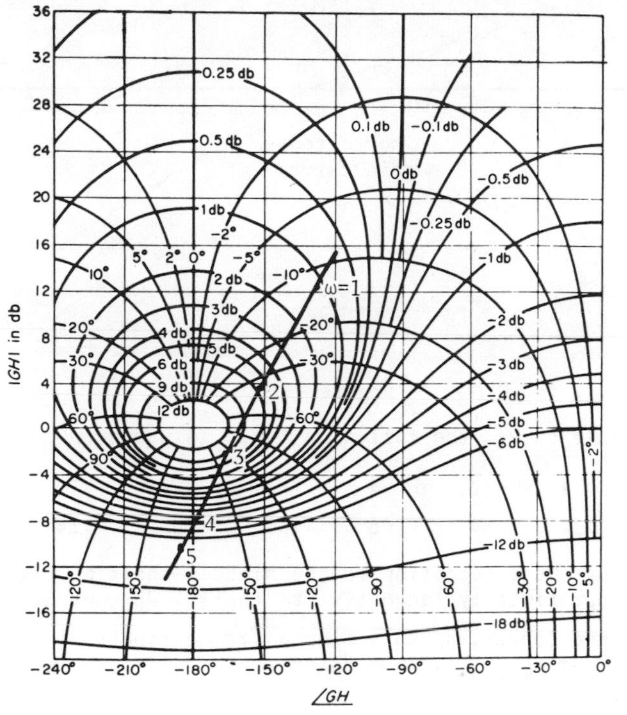

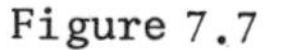
Figure 7.7

Example 7.3

The open-loop transfer function of a unity feedback system is given as

$$G(j\omega) = \frac{50}{(1 + j\omega 0.05)(1 + j\omega 2.5)}$$

Draw a corrected Bode diagram and hence by transferring the relevant information to a Nichols chart, determine the gain margin, phase margin and M_{max}. Hence find the damping ratio and the natural undamped frequency of the system.

For a Bode diagram, shown in fig. 7.8, there are two corner frequencies 0.4 rad s^{-1} and 20 rad s^{-1}, while the calculation for the gain of 50 is

$$20 \log_{10} 50 = 34 \text{ dB}$$

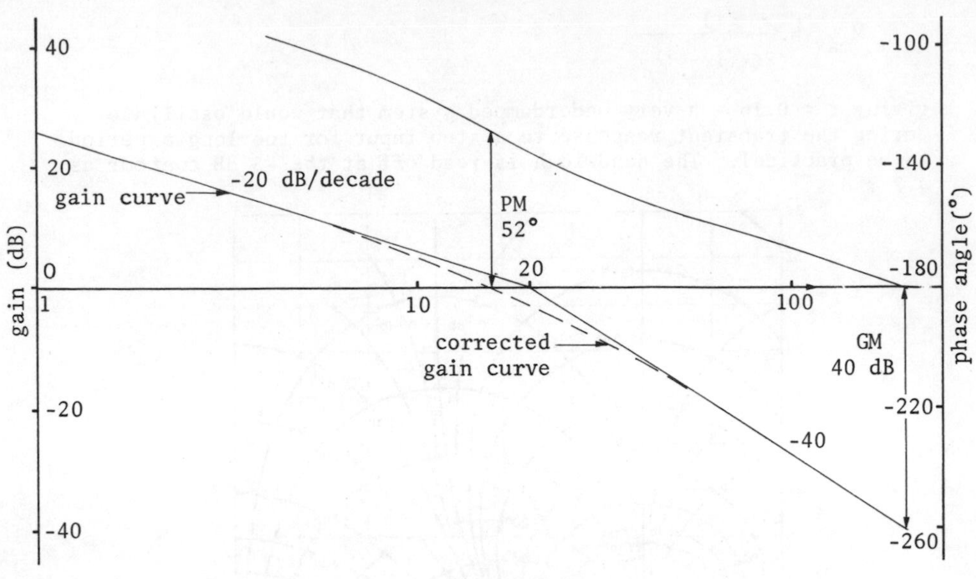

Figure 7.8

For the phase curve, the following table is prepared.

ω rad s^{-1}	1	2	4	6	8	10	20	40
$\underline{|G(j\omega)}$°	-11	$-84\frac{1}{2}$	-96	-102	-109	-114	-134	-152

Transferring the information to the Nichols chart of fig. 7.9, the locus is approximately tangential to the 1.1 dB contour, i.e. M_{max} = 1.1 dB, and converting this figure from decibels yields $20 \log_{10} M = 1.1$ dB, or $M_{max} = 1.135$. Using

$$M = \frac{1}{2\zeta\sqrt{(1 - \zeta^2)}}$$

yields $\zeta = 0.513$.

The resonant frequency ω can be interpolated as approximately 15.5 rad s^{-1}, so that from $\omega_r = \omega_n\sqrt{(1 - 2\zeta^2)}$, calculate $\omega_n \simeq 22$ rad s^{-1}. Since the open-loop transfer function for this system is of the second-order, a direct comparison for these answers can be made.

The characteristic equation of the system is

$$s^2 + 20.4s + 408 = 0$$

to be compared with

$$s^2 + 2\zeta\omega_n s + \omega_n^2 = 0$$

yielding $\omega_n = 20.2$ rad s^{-1} and $\zeta = 0.5$.

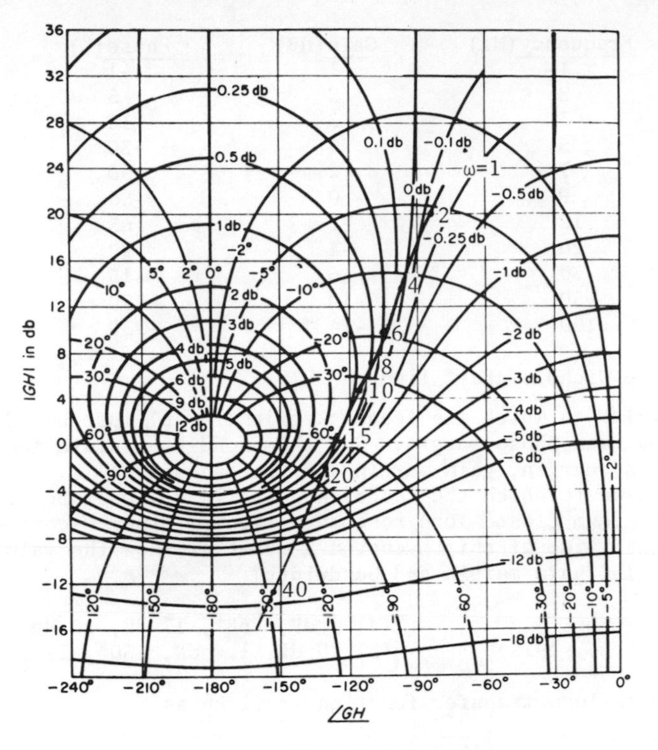

Figure 7.9

These answers can be said to show up errors in Bode and Nichols chart usage but for design purposes they are sufficiently accurate for initial investigations. Large-scale Nichols charts would obviously improve the accuracy of the graphical results.

Again, if required the bandwidth can be read off as approximately 25 rad s^{-1}.

Problems

1. An open-loop transfer function is given as

$$G(s) = \frac{40}{s(1 + 0.0625s)(1 + 0.25s)}$$

and is found to give an unstable system in closed-loop. A network whose transfer function is (1 + 4s)/(1 + 80s) is added in the forward path of the system. Find from a Nichols chart the gain margin, the phase margin, M_{max} and the corresponding resonant frequency as well as the system bandwidth.

[20 dB, 52°, 1.4 dB, 0.9 rad s^{-1}, 3 rad s^{-1}]

2. The closed-loop frequency response of a system with unity negative feedback gain is

Frequency (Hz)	Gain (dB)	Phase (°)
1	1	-1
2	2	-5
3	3	-10
5	4	-30
7	2	-50
9	0	-55
12	-2	-65
20	-4	-90
30	-6	-115
50	-10	-170
100	-14	-230

Using the Nichols chart find

(a) the value to which the feedback gain (in dB) must be adjusted to give maximum phase margin. For this condition state the values of the phase margin, gain margin and the bandwidth.
(b) the value to which the feedback gain (in dB) must be adjusted to give minimum closed-loop resonance (i.e. minimum M-contour). What is the value of this M-contour? What are now the values of phase margin, gain margin and bandwidth?

(C.E.I. C.S.E., 1975)

[(a) 2 dB, -68°, 12 dB, 20 Hz
(b) 10 dB, 1.4 dB, -30°, 25 dB, 100 Hz]

3. The open-loop transfer function is given as

$$G(s) = \frac{10}{s(1 + 0.2s)(1 + 0.02s)}$$

Determine the dB magnitude-phase angle locus and from it, find

(a) the gain and phase margins
(b) M_{max} and the resonant frequency.

[(a) 9.5 dB, 25°,
(b) 8 dB, 7.2 rad s^{-1}]

8 THE ROOT LOCUS PLOT

The basic characteristic of the transient response of a closed-loop system is determined from the closed-loop poles. In the design of closed-loop systems, it is usually necessary to adjust the open-loop poles and zeros so as to place the closed-loop poles at desirable locations in the s plane.

The closed-loop poles are the roots of the characteristic equation $1 + G(s)H(s) = 0$, and the equation is satisfied when $|G(s)H(s)| = 1$ and $\underline{/G(s)H(s)} = 180°$. The process of finding points on the s plane that satisfy these conditions is the basis of the root locus method, i.e. the roots of the characteristic equation are plotted for all values of a system parameter, which is usually but not necessarily the system gain.

Consider a unity feedback control system with an open-loop transfer function

$$G(s) = \frac{K}{s(s + 2)}$$

The characteristic equation is

$$s^2 + 2s + K = 0$$

and the two roots s_1 and s_2 are $-1 \pm \sqrt{(1 - K)}$. These are shown plotted in fig. 8.1 as a function of K on the s plane. This plotting technique is the heart of the root locus method.

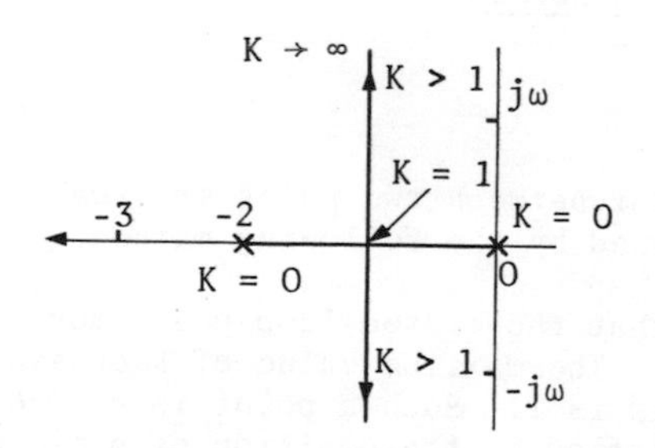

Figure 8.1

This example is simple enough to illustrate the method of calculating the position of the closed-loop poles without reference to the following rules that are available for more complex equations.

Rule 1

The number of loci is equal to the order of the characteristic equation.

Rule 2

Each locus starts at an open-loop pole when $K = 0$ and finishes either at an open-loop zero or infinity, when $K = \infty$.

Rule 3

Loci either move along the real axis or occur as complex conjugate pairs of loci.

Rule 4

A point on the real axis is part of the locus if the number of poles and zeros to the right of the point concerned is odd.

Rule 5

When the locus is far enough from the open-loop poles and zeros, it becomes asymptotic to lines making angles to the real axis, given by

$$\pm \frac{n\pi}{P - Z}$$

where P is the number of open-loop poles, Z the number of open-loop zeros and n is an odd integer.

Rule 6

The asymptotes intersect the real axis at a point x given by

$$x = \frac{\Sigma \text{ poles} - \Sigma \text{ zeros}}{P - Z}$$

Rule 7

The break-away point between two poles or break-in point between two zeros can be obtained by the following method.

Fig. 8.1 shows that the closed-loop poles move along the real axis as K is increased. The maximum value of K obtained before the poles leave the real axis is 1. Such a point is called a break-away point. The condition satisfied by the position of a closed-loop pole is $KG(s) = -1$.

It is convenient to invert G(s) and present the equation as $K/V(s) = -1$ or $V(s) = -K$. To investigate the variation of K with complex frequency, and in particular the maximum value of K for real values of s, differentiate V(s) with respect to s and equate to zero.

Rule 8

The limiting value of K for stability may be found using the Routh array on the characteristic equation and hence the value of the loci at the intersection with the imaginary axis is determined.

Rule 9

The angle of departure from a complex pole or zero is given by

$$\text{Angle of departure} = 180° - \phi_P + \phi_Z$$

where ϕ_P is the sum of all the angles subtended by the other poles and ϕ_Z is the sum of the angles of any zeros.

These rules are sufficient to establish the main outline of the loci. The method used to determine intermediate points on the loci or values of K at specific positions will be illustrated in several of the examples in this chapter.

Appendix C shows sketches of the more common loci up to three poles and no zeros.

Example 8.1

Consider a control system with a forward transfer function

$$G(s) = \frac{K}{s(s + 1)(s + 2)}$$

and $H(s) = 1$

Sketch the root locus plot and then determine the value of K so that the damping ratio is approximately 0.5.

This system has three open-loop poles at 0, -1 and -2. The characteristic equation is a cubic, so there will be three sections of loci and, since there are no zeros, all loci move to infinity.

$$s^3 + 3s^2 + 2s + K = 0$$

Since all three poles are real, parts of the negative real axis will be loci - rule 4. The angles of the asymptotes are obtained from $n\pi/3$, i.e. 60°, 180° and 300°, corresponding to n = 1, 3 and 5 - rule 5. The intersection with the real axis is obtained from

$$\frac{0 - 1 - 2}{3} = -1$$

rule 6. The break-away point is found by differentiating V(s) with respect to s - rule 7.

$$V(s) = \frac{1}{G(s)} = s^3 + 3s^2 + 2s$$

$$\frac{dV}{ds} = 3s^2 + 6s + 2 = 0$$

$$s_b = -1.577 \quad \text{or} \quad -0.423$$

Since break-away point lies between 0 and -1 the value to s_b is -0.423.

From the characteristic equation

$$s^3 + 3s^2 + 2s + K = 0$$

by using the Routh array the limiting value of K for stability is found.

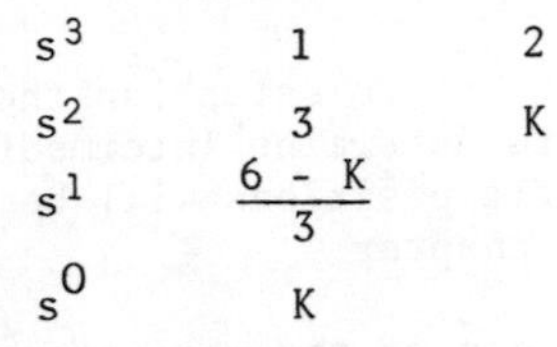

s^3	1	2
s^2	3	K
s^1	$\frac{6 - K}{3}$	
s^0	K	

The limiting value of K is 6, thus the intersection with the imaginary axis is found from

$$3s^2 + 6 = 0$$

or $\quad s = \pm j\sqrt{2}$

rule 8. The general form of the locus can now be sketched as in fig. 8.2.

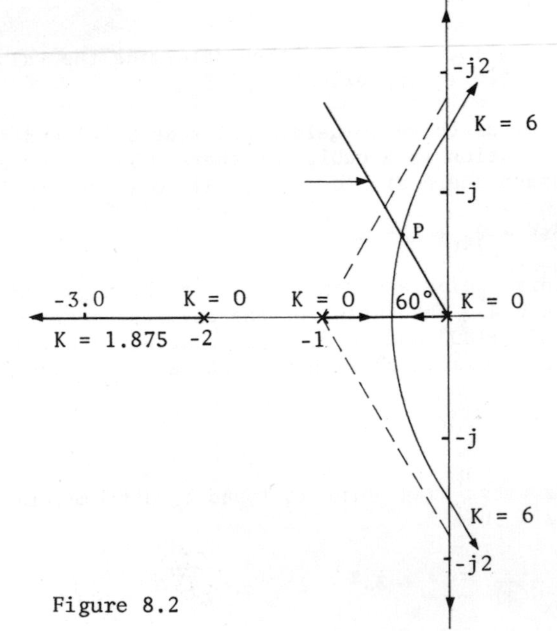

Figure 8.2

Intermediate points may be determined by the following method, since $|G(s)| = 1$.

At $s = -2.5$

$$K = 0.5 \times 1.5 \times 2.5 = 1\tfrac{7}{8} = 1.875$$

$$\begin{array}{rl}
s + 2.5 & \underline{)s^3 + 3s^2 + 2s + 1.875}(s^2 + 0.5s + 0.75 \\
 & \underline{s^3 + 2.5s^2} \\
 & \qquad 0.5s^2 + 2s \\
 & \qquad \underline{0.5s^2 + 1.25s} \\
 & \qquad\qquad 0.75s + 1.875 \\
 & \qquad\qquad \underline{0.75s + 1.875}
\end{array}$$

By means of this division, the other two roots of the cubic equation can be found. The roots of the quadratic equation $s^2 + 0.5s + 0.75 = 0$ yield $s = -0.25 \pm j0.82$ and these are the other two points on the loci where $K = 1.875$.

For the final part of the question, draw a line from the origin that makes an angle of 60° with the real axis, as shown in fig. 8.2, i.e. the damping ratio is 0.5 and $\cos^{-1} 0.5$ is 60°. Then evaluate the value for K at the intersection point P, by measuring or calculating the distance from each pole to the point P. The co-ordinates of P are $-0.33 + j0.58$, i.e.

$$K = 0.667 \times 0.908 \times 1.768 = 1.07$$

Example 8.2

Consider the control system with a forward transfer function

$$G(s) = \frac{K(s + 2)}{s^2 + 2s + 3}$$

and $H(s) = 1$

Sketch the root locus plot and determine the approximate damping ratio for a value of $K = 1.33$.

It is seen that $G(s)$ has a pair of complex conjugate poles at $s = -1 \pm j\sqrt{2}$ and one zero at -2. The characteristic equation is a quadratic

$$s^2 + s(2 + K) + 3 + 2K = 0$$

so there will be two sections of loci, each of which starts from one of the two complex conjugate poles and breaks in the part of the negative real axis between -2 and $-\infty$.

Since there are two open-loop poles and one zero there is one asymptote, which coincides with the negative real axis.

The determination of the angle of departure of the loci from the open-loop poles is found by the use of rule 9

Angle of departure = 180° - 90° + 55° = 145°

The break-in point is found using rule 7

$$V(s) = \frac{s^2 + 2s + 3}{s + 2}$$

$$\frac{dV}{ds} = \frac{s^2 + 4s + 1}{(s + 2)^2} = 0$$

Solution of $s^2 + 4s + 1 = 0$ yields -3.73 as the break-in point.

The locus is shown in fig. 8.3. To find the point on the locus where K = 1.33 evaluate as follows

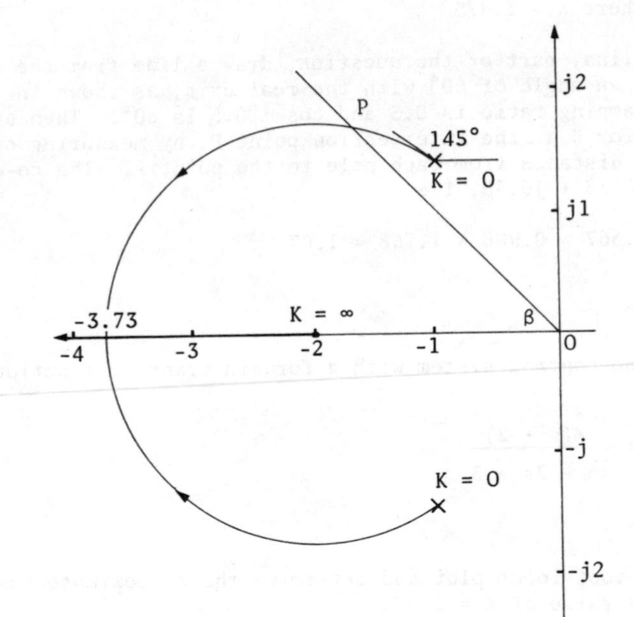

Figure 8.3

$$\frac{1.33(s + 2)}{s^2 + 2s + 3} = - 1$$

or $s^2 + 3.33s + 5.66 = 0$

therefore

$$s = - 1.665 \pm j1.699$$

This point is shown in fig. 8.3 at P. Now draw a line from the origin to P and, from the angle β , the damping ratio is cos 45° or 0.7.

Example 8.3

Sketch the root locus of the following systems

(a) $\dfrac{K}{s^2 + 10s + 100}$

(b) $\dfrac{K(s + 1)}{s^2 + s + 10}$

(c) $\dfrac{K(s + 1)}{s^2 (s + 2)}$

In all cases, only compute the information necessary to enable the loci to be sketched.

(a) $s^2 + 10s + 100 = 0$

$$s_{1,2} = - 5 \pm j8.66$$

there are two loci with asymptote angles of $n\pi/2$, i.e. 90° and 270°. The intersection of these asymptotes is (- 5 - 5)/2 or -5. The angle of departure from these complex poles is given by (180° ± 90°) or 90° and 270°. The loci are shown in fig. 8.4a.

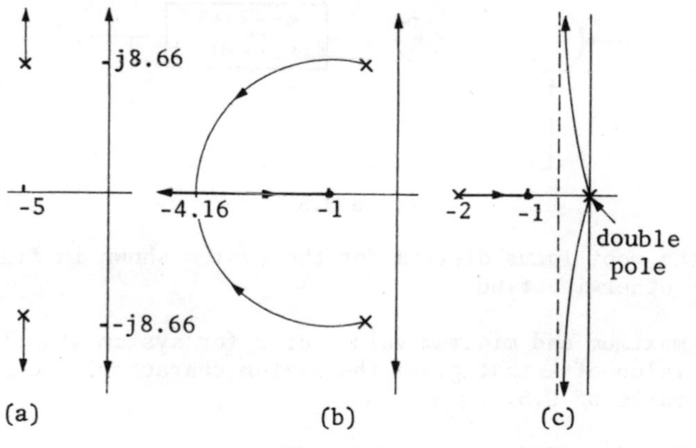

Figure 8.4

(b) $s^2 + s + 10 = 0$

$$s_{1,2} = - 0.5 \pm j3.12$$

there are two complex poles and one zero, but two loci. Asymptote angle is 180°/(2 - 1) or 180°, i.e. the negative real axis.

For the break-in point differentiate V(s) with respect to s.

$$V(s) = \frac{s^2 + s + 10}{s + 1}$$

$$\frac{dV}{ds} = \frac{s^2 + 2s - 9}{(s + 1)^2} = 0$$

therefore

$$s_b = - \frac{2 \pm \sqrt{(4 + 36)}}{2} = - 4.16 \text{ or } + 2.16$$

The break-in point is -4.16

The locus from the complex poles is part of a circle, centre -1, and is shown in fig. 8.4b.

(c) This system is a third-order and therefore has three loci; the asymptote angles are $n\pi/(3 - 1)$ or 90°, and 270° corresponding to n = 1 and 3; the real axis forms the line of the third locus. The intersection of the asymptotes is

$$\frac{-2 - (-1)}{3 - 1} = \frac{-1}{2}$$

The loci are shown in fig. 8.4c.

Example 8.4

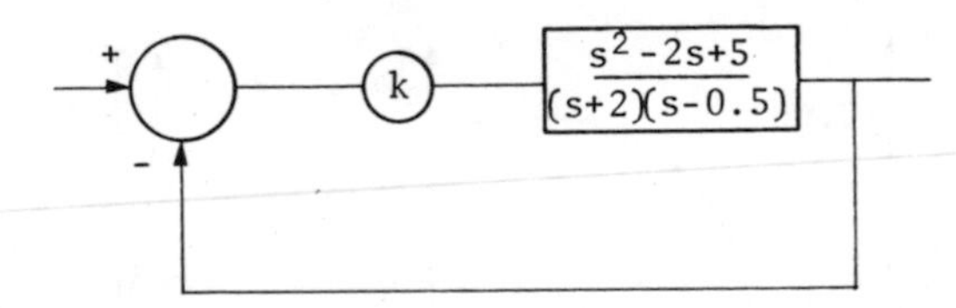

Figure 8.5

Sketch the root locus diagram for the system shown in fig. 8.5. Hence or otherwise find

(a) the maximum and minimum values of k for system stability, and
(b) the value of k that gives the system characteristic equation a damping ratio of 0.5.

(C.E.I. C.S.E., 1974)

From an inspection of the forward transfer function it is evident that this system is of the non-minimum phase type, i.e. the open-loop zeros are complex ($1 \pm j2$) but with positive real terms, and one of the poles is also positive.

This means that part of the root locus will lie in the right-hand half of the s plane.

$$G(s) = \frac{k(s^2 - 2s + 5)}{s^2 + 1.5s - 1}$$

The characteristic equation is $s^2(1 + k) + s(1.5 - 2k) + 5k - 1 = 0$. The Routh array is

$$\begin{array}{lll} s^2 & 1 + k & 5k - 1 \\ s^1 & 1.5 - 2k & \\ s^0 & 5k - 1 & \end{array}$$

(a) $2k = 1.5 \quad k_{max} = 0.75$

$5k = 1 \quad k_{min} = 0.2$

To find the break-point for the locus, differentiate V(s) with respect to s and equate to zero.

$$V(s) = \frac{s^2 + 1.5s - 1}{s^2 - 2s + 5}$$

$$\frac{dV}{ds} = \frac{-3.5s^2 + 12s + 5.5}{(s^2 - 2s + 5)^2} = 0$$

therefore

$$s_b = + 3.84 \text{ or } - 0.41$$

Calculation of k at the break-point yields

$$k_b = \frac{0.91 \times 1.59}{5.988} = 0.242$$

when $k = 0.75$

$$1.75s^2 + 2.75 = 0$$

$$s = \pm j1.25$$

when $k = 0.2$

$$1.2s^2 = 0$$

i.e. the origin. The loci are shown in fig. 8.6.

(b) For $\zeta = 0.5$ draw a line at an angle of 60° as shown and calculate k at the intersection between loci and line.

$$k = \frac{1.85 \times 1.05}{1.95 \times 2.95} = 0.338$$

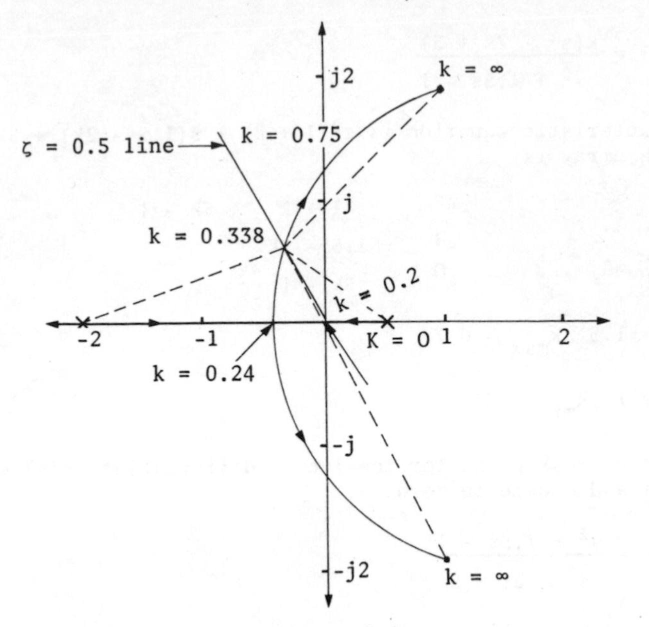

Figure 8.6

Note on Non-minimum Phase Systems

The definition of a minimum phase network or system is one whose phase shift is the minimum possible for the number of energy-storage elements in the system. This definition restricts the zeros and poles of minimum phase systems to the left-hand half of the complex s plane.

If a system has one pole or zero in the right-hand half of the s plane, then the system is called non-minimum phase. The term 'minimum phase' comes from the characteristic of a phase change in such a system when subject to a sinusoidal input.

Example 8.5

Consider the unity feedback system whose forward transfer function is given as

$$G(s) = \frac{10(s + 1)}{s(s - 3)}$$

Explain why this system has an unstable feed forward transfer function. Sketch the root locus and locate the closed-loop poles. Show that although the closed-loop poles lie on the negative real axis and the system is not oscillatory, the unit step response will exhibit an overshoot.

The root locus for this system is shown in fig. 8.7 with open-loop poles at s = 0 and +3, i.e. unstable characteristic, and a zero at -1.

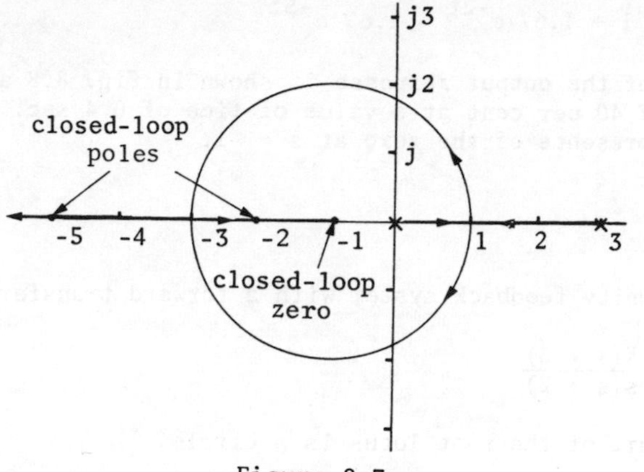

Figure 8.7

The closed-loop transfer function is

$$\frac{\theta_o}{\theta_i} = \frac{10(s + 1)}{s^2 + 7s + 10} = \frac{10(s + 1)}{(s + 5)(s + 2)}$$

i.e. two poles at -5 and -2.

For a unit step input

$$\theta_o(s) = \frac{10(s + 1)}{s(s + 2)(s + 5)}$$

Solving via Laplace and the use of partial fractions

$$\frac{A}{s} + \frac{B}{s + 2} + \frac{C}{s + 5} \equiv \frac{10(s + 1)}{s(s + 2)(s + 5)}$$

$$A(s^2 + 7s + 10) + Bs(s + 5) + Cs(s + 2) \equiv 10(s + 1)$$

therefore A = 1, B = + 1.67 and C = - 2.67. Thus

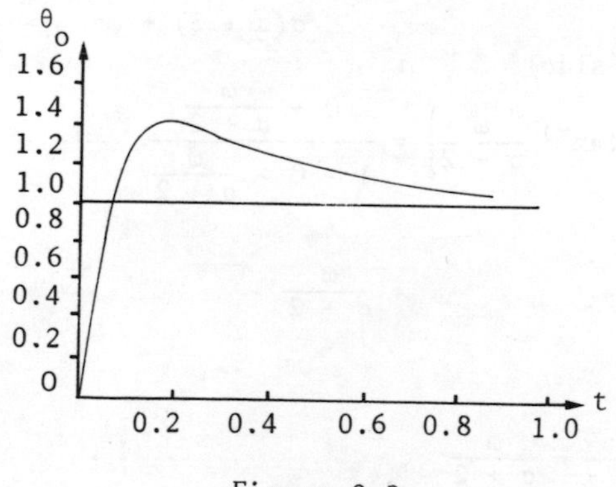

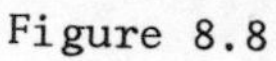

Figure 8.8

$$\theta_o(t) = 1 + 1.67\, e^{-2t} - 2.67\, e^{-5t}$$

The sketch of the output response is shown in fig. 8.8 and shows overshoot of 40 per cent at a value of time of 0.4 sec. This is due to the presence of the zero at $s = -1$.

Example 8.6

Consider a unity feedback system with a forward transfer function

$$G(s) = \frac{K(s + 3)}{s(s + 2)}$$

Show that part of the root locus is a circle.

The angle condition can be applied in the following way

$$\angle G(s) = \angle s + 3 - \angle s - \angle s + 2 = 180°$$

Now s is complex and may be represented as $\sigma + j\omega$. Hence

$$\angle \sigma + j\omega + 3 - \angle \sigma + j\omega - \angle \sigma + j\omega + 2 = 180°$$

and this can be written as

$$\tan^{-1} \frac{\omega}{\sigma + 3} - \tan^{-1} \frac{\omega}{\sigma} = 180° + \tan^{-1} \frac{\omega}{\sigma + 2}$$

Now take the tangents of both sides

(left-hand side)

$$\tan \left(\tan^{-1} \frac{\omega}{\sigma + 3} - \tan^{-1} \frac{\omega}{\sigma}\right) = \frac{\dfrac{\omega}{\sigma + 3} - \dfrac{\omega}{\sigma}}{1 + \left(\dfrac{\omega}{\sigma + 3}\right)\left(\dfrac{\omega}{\sigma}\right)}$$

$$= \frac{-\,3\omega}{\sigma(\sigma + 3) + \omega^2}$$

(right-hand side)

$$\tan \left(180 + \tan^{-1} \frac{\omega}{\sigma + 2}\right) = \frac{0 + \dfrac{\omega}{\sigma + 2}}{1 - 0 \times \dfrac{\omega}{\sigma + 2}}$$

$$= \frac{\omega}{\sigma + 2}$$

Hence

$$\frac{-\,3\omega}{\sigma(\sigma + 3) + \omega^2} = \frac{\omega}{\sigma + 2}$$

or $(\sigma + 3)^2 + \omega^2 = (\sqrt{3})^2$

This equation represents a circle with centre at $\sigma = -3$, $\omega = 0$ and with radius $\sqrt{3}$.

Note that the centre of the circle is the zero of the open-loop transfer function.

Example 8.7

Construct the root locus of the open-loop transfer function of example 8.6. Hence determine the damping ratio for maximum oscillatory response. What is the value of K at this point of the locus? Determine the unit-step response for such a unity feedback system.

For this transfer function, the solution to example 8.6 yields the fact that part of the root locus is circular, i.e. centre at $s = -3 + j0$ and radius $\sqrt{3}$. The complete locus is shown in fig. 8.9.

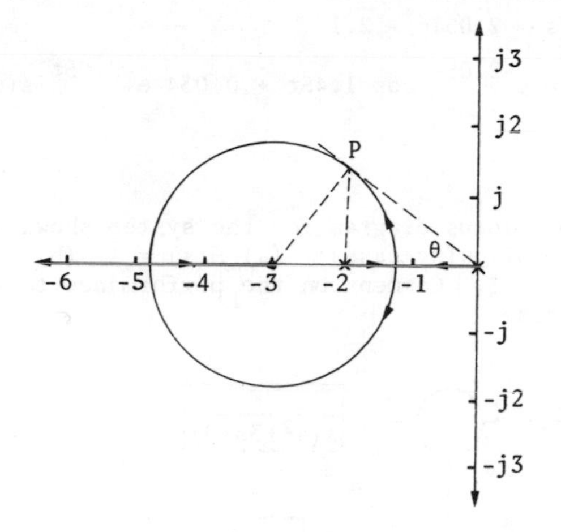

Figure 8.9

Next draw a line from the origin tangential to the circular part of the locus, point P as shown. The angle θ is measured as 34°. Hence the damping ratio is

$$\zeta = \cos\theta = 0.82$$

The value of K can be found in two ways. The two closed-loop poles are $-2.05 \pm j1.45$. Thus from the second-order characteristic equation

$$s^2 + s(2 + K) + 3K \equiv (s + 2.05 + j1.45)(s + 2.05 - j145)$$

therefore

$$K = 2.1$$

The alternative method is to measure, from the diagram, the distance from each open-loop pole and zero to the point P. Thus

$$K = \frac{2.45 \times 1.48}{1.73} = 2.1$$

The closed-loop transfer function of this system is

$$\frac{\theta_o}{\theta_i} = \frac{2.10(s + 3)}{s^2 + 4.10s + 6.3}$$

For $\theta_i = \frac{1}{s}$

$$\theta_o = \frac{2.1(s + 3)}{s((s + 2.05)^2 + 2.1)}$$

$$\theta_o = \frac{1}{s} - \frac{(s + 2.05 - 0.05)}{(s + 2.05)^2 + 2.1}$$

$$\theta_o(t) = 1 - e^{-2.05t} \cos 1.45t + 0.034\, e^{-2.05t} \sin 1.45t$$

Problems

1. Sketch a root locus diagram for the system shown in fig. 8.10 for each of the following cases: (a) $H(s) = 1$, (b) $H(s) = (s + 1)$ and (c) $H(s) = s + 3$. Comment on the performance that would be expected in each case.

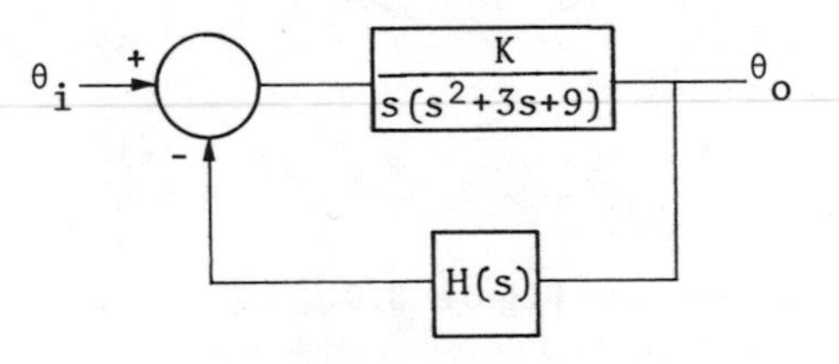

Figure 8.10

2. A block diagram of a servo mechanism is shown in fig. 8.11. Sketch the root loci for the system and indicate the maximum value of K that can be used without instabilities arising.

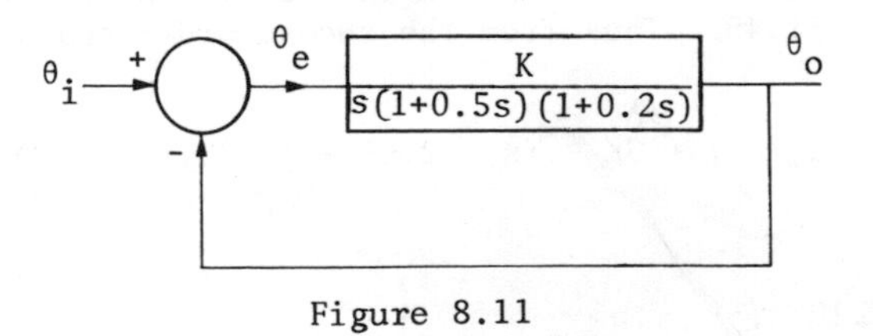

Figure 8.11

How can the frequency response of the system be estimated for any fixed gain value?

(C.E.I. S & C Eng., 1970) [$K < 7$]

3. The open-loop transfer function of a control system is of the form

$$G(s) = \frac{K(s + a)}{s^2(s + b)}$$

Show the way in which the root loci change as a varies over the range from zero to b.

If a = 2 and b = 10, sketch the root locus and determine approximately the value of K that gives the greatest damping ratio for the oscillatory mode.

[44.6]

4. Sketch the root locus of the system whose open-loop transfer function is

$$G(s) = \frac{K}{s(s + 2)(s + 7)}$$

Hence deduce (a) the maximum value of K for stability, and (b) the value of K that provides a damping ratio of 0.7.

[$K < 126$; $K \simeq 11.5$]

5. Write short notes on the rules applied to root locus plotting. Describe how you would (a) determine the gain at any point on the root locus and (b) determine the frequency response from the pole zero positions.

(C.E.I. S & C Eng., 1969)

6. Sketch the root locus for the system shown in fig. 8.12. From the root locus diagram determine a value of gain K for which the system becomes unstable.

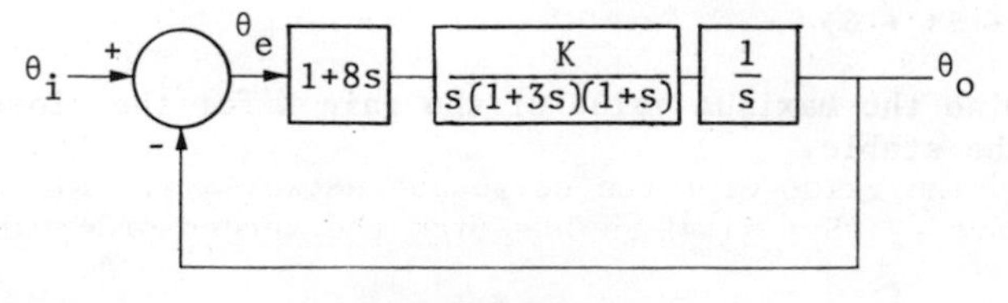

Figure 8.12

How can the frequency response of the system be estimated for any fixed gain value?

(C.E.I. S & C Eng., 1968) [$K > 1/12$]

7. Draw the root locus diagram for a control system with unity feedback having the forward transfer function

$$\frac{K}{s(s + 2)(s + 5)}$$

Give all relevant characteristics of the curves that are useful in establishing the locus.

Find the least value of K to give an oscillatory response and the greatest value of K that can be used before continuous oscillation occurs. Find the frequency of the continuous oscillation when K is just large enough to give this condition.

[$K > 4.06$, $K < 70$, 3.16 rad s^{-1}]

8. A system has a forward path transfer function

$$\frac{K}{s(s + 1)}$$

It is required that the closed-loop poles be located at $s = -1.6 \pm j4$ using a phase-advance network with transfer function

$$\frac{(s + 2.5)}{(s + \alpha)}$$

as compensation in the forward path. Determine (a) the required value of α, (b) the value of K to locate the closed-loop poles as required and (c) the location of the third closed-loop pole.

[$\alpha = 5.3$; 23.2; $s = -3.1$]

9. Discuss the usefulness of the root locus method.

Sketch the root locus for the system with a forward path transfer function

$$\frac{K}{s(s^2 + 5s + 6)}$$

and determine the maximum value of the gain K for the closed-loop system to be stable.

[$K = 30$]

9 MISCELLANEOUS PROBLEMS

1. The temperature of a plastic moulding process is controlled by heating the jacket surrounding the plastic charge (fig. 9.1). The proportional controller supplies heat equal to B times the temperature error (in K) at the jacket. The jacket temperature changes by 2×10^{-6} K J^{-1} of heat supplied or extracted. The temperature C of the plastic charge changes by 0.004 (K s^{-1})/K temperature difference from the jacket. The temperature difference in the jacket may be assumed uniform.

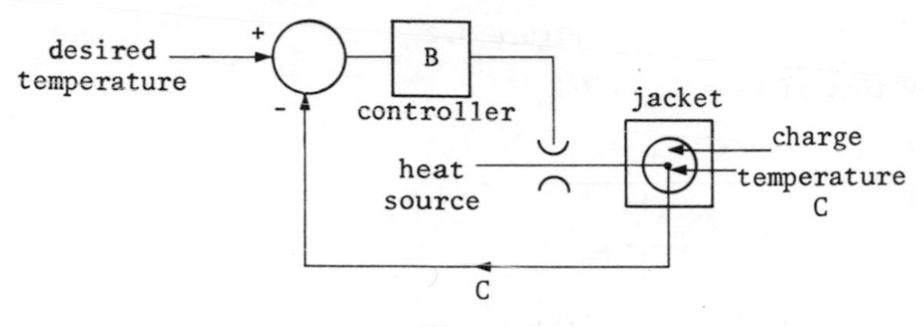

Figure 9.1

(a) Determine the controller gain B for a step response with a damping ratio of 0.7.

(b) Show how the speed of response can be increased while maintaining the same degree of damping.

(C.E.I. S & C Eng., 1972) [1.02×10^3J s^{-1} K]

2. An electric motor whose armature has a moment of inertia 0.1 kg m^2 drives an inertia load of 90 kg m^2 through frictionless gears. The motor torque can be considered independent of speed, having value 20 N m. Find the gear ratio that gives maximum load acceleration, and the time then taken to reach a speed of 50 rad s^{-1} from rest. Derive all formulae used.

(C.E.I. S & C Eng., 1974) [30:1; 0.5 s]

3. The forward path transfer function of a unity feedback control system is

$$G(s) = \frac{3(s + 1)}{s^2(2s + 1)}$$

(a) Show that the system is unstable.

(b) Show that negative velocity feedback cannot be used for stabilisation if the maximum allowable steady-state error for a unit ramp input is 0.1.

(C.E.I. S & C Eng., 1973)

4. The bridged-T network shown in fig. 9.2 is often known as a rejection network. Why is this so?

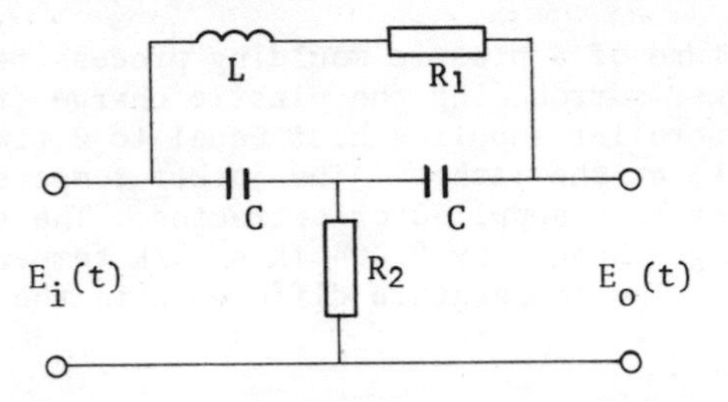

Figure 9.2

Show that if $R_2 = (\omega_n L)^2/4R_1$

$$\frac{E_o}{E_i}(s) = \frac{s^2 + \omega_n^2}{s^2 + \frac{2\omega_n s}{K} + \omega_n^2}$$

where $\omega_n^2 = 2/(LC)$, $K = \omega_n L/R_1$.

(C.E.I. S & C Eng., 1973)

5. (a) Discuss the use of linearisation techniques in the analysis and design of non-linear systems.

(b) Fig. 9.3 shows a hydraulic relay consisting of a flapper nozzle valve and a single acting hydraulic cylinder. The piston rod drives a load of mass M and spring rate k. Friction is negligible.

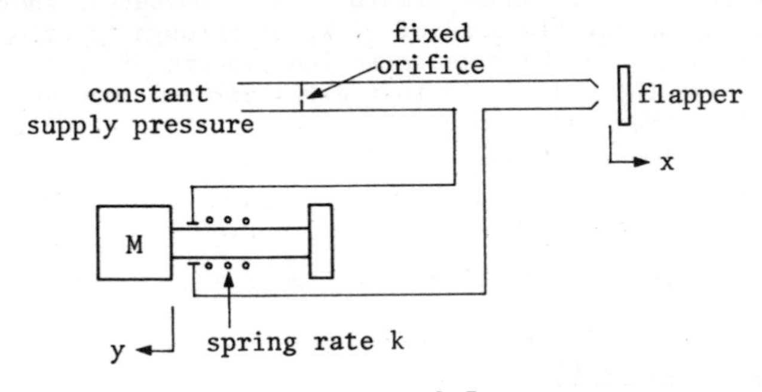

Figure 9.3

(a) Derive the linearised system equations, neglecting oil compressibility.

(b) Hence derive the transfer function relating small changes in flapper displacement x and load displacement y.

(C.E.I. S & C Eng., 1972)
$$\frac{k_F A/(k_o + k_n)}{Ms^2 + \dfrac{A^2 s}{k_o + k_n} + k}$$

6. Determine the transfer function for the network shown in fig. 9.4. You may assume that the source impedance is zero and that the load impedance is infinite.

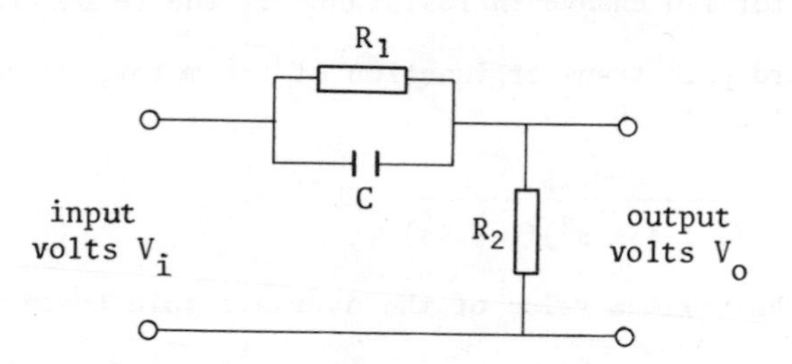

Figure 9.4

If the resistance R_2 is known, show how the values of resistance R_1 and capacitance C can be found from (a) the step response and (b) the frequency response.

(C.E.I. S & C Eng., 1971)
$$\frac{\alpha(1 + sT)}{1 + s\alpha T};\ \alpha = \frac{R_2}{R_1 + R_2};\ T = CR_1$$

7. The block diagram of an attitude stabilising system for a VTOL aircraft is shown in fig. 9.5. If the aircraft dynamics are represented by the transfer function $G_2(s) = 10/(s^2 + 0.25)$, the transfer function of the filter and actuator is

$$G_1(s) = \frac{6(s + 7)}{s + 2}$$

the transfer function of the rate gyro may be taken as $H(s) = s$.

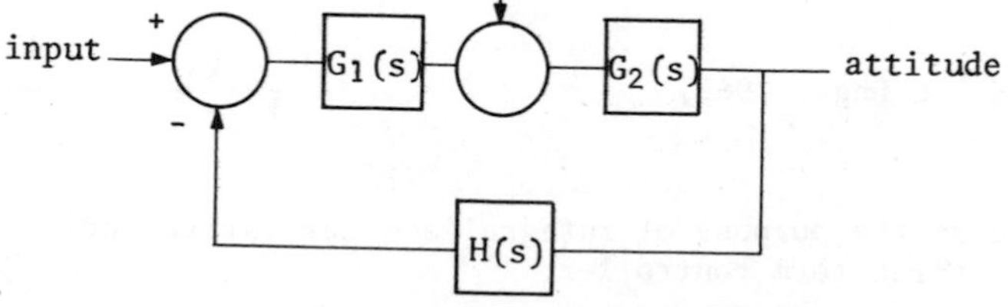

Figure 9.5

(a) Find the system phase margin.

(b) Find the steady-state attitude error resulting from a step wind disturbance.

(C.E.I. S & C Eng., 1971) [88°, 0.017]

8. The main kettle in a plastics plant is operated by a gas-fired heating system. The gas-control valve is operated by an electric motor, which in turn receives its actuating signal from a resistance-type thermometer. The thermometer is situated in a leg of a bridge circuit. The nominal resistance of each arm of the bridge is 100 Ω and the bridge supply is 40 V. Determine the voltage change across the bridge for 1 Ω change in resistance of the temperature element.

The forward path transfer function of the motor, valve and temperature measuring element is

$$KG(s) = \frac{0.25}{(1 + s + s^2)(1 + 4s)} \quad \Omega V^{-1}$$

Determine the maximum value of the over-all gain to ensure stability.

(C.E.I. S & C Eng., 1970) [0.0995 to -0.1005 V; 5.25]

9. A mechanical system reduces to the arrangement shown in fig. 9.6. This consists of two springs of stiffness k_1 and k_2 and a dashpot of damping coefficient δ. The lower spring of stiffness k_2 is fixed at one end, while the input and output displacements are defined as x and y.

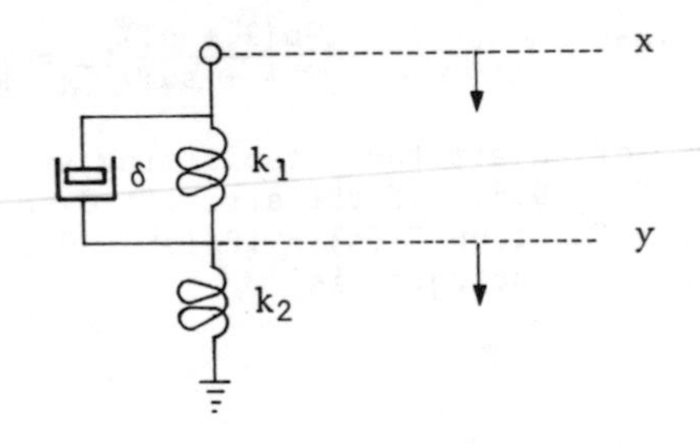

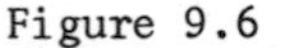
Figure 9.6

Derive the transfer function for this arrangement and develop an equivalent electrical passive network giving an analogous dynamic response.

(C.E.I. S & C Eng., 1969)

$$\frac{k_1}{k_1 + k_2} \times \frac{1 + \dfrac{\delta}{k_1}s}{1 + \dfrac{\delta}{k_1 + k_2}s}$$

10. Discuss the purpose of integral and derivative action in relation to a three-term controller.

A process has the following elements

(a) a three-term controller whose gain is set at 5

(b) a process whose transfer function can be considered as a single exponential lag of time constant 4 min
(c) a measuring unit whose transfer function can also be considered as a simple exponential lag of time constant 2 min

Determine the minimum value of the integral action time to ensure stability for all settings of the derivative action. Unity feedback can be assumed.

(C.E.I. S & C Eng., 1969) [10/9]

11. A control system has unity feedback and the following forward path transfer function

$$G(s) = \frac{K}{s(1 + 0.2s)}$$

Design a phase-lead compensation network to give a velocity error coefficient of at least 100 s^{-1} and a phase margin of at least 45°.

12. A control system with unity and derivative feedback is shown in fig. 9.7. If the feedback is zero determine the damping ratio and natural frequency of the system and the steady-state error for unity ramp input. A damping ratio of 0.8 is obtained by adjusting the derivative feedback. Determine the value of the constant k to achieve this condition and estimate the increase in the steady-state error of the system for unity ramp input.

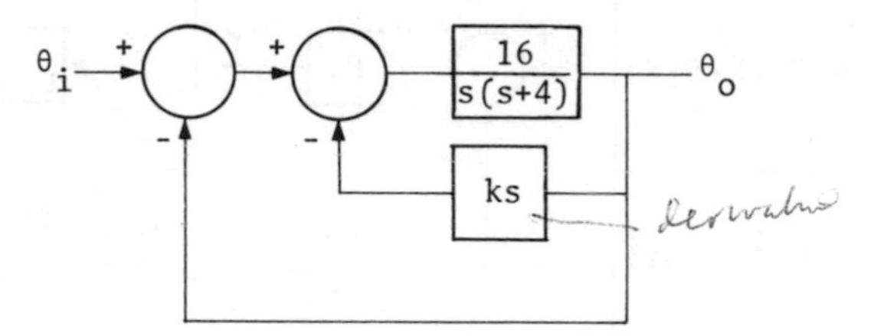

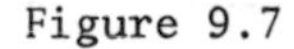

Figure 9.7

[ζ = 0.5, ω_n = 4, 0.25; k = 0.15, 0.4]

APPENDIX A Block diagram theorems

	Transformation	Equation	Block Diagram	Equivalent Block Diagram
1	Combining blocks in cascade	$Y = (P_1P_2)X$	X → P_1 → P_2 → Y	X → P_1P_2 → Y
2	Combining blocks in parallel; or eliminating a forward loop	$Y = P_1X \pm P_2X$	X → P_1 (+), P_2 (±) → summing junction → Y	X → $P_1 \pm P_2$ → Y
3	Removing a block from a forward path	$Y = P_1X \pm P_2X$		X → P_2 → $\frac{P_1}{P_2}$ (+), feedforward (±) → summing junction → Y
4	Eliminating a feedback loop	$Y = P_1(X \mp P_2Y)$	X (+) → summing junction → P_1 → Y; Y → P_2 → (∓)	X → $\frac{P_1}{1 \pm P_1P_2}$ → Y
5	Removing a block from a feedback loop	$Y = P_1(X \mp P_2Y)$		X → $\frac{1}{P_2}$ (+) → summing junction → P_1P_2 → Y; feedback (∓)

	Transformation	Equation	Block Diagram	Equivalent Block Diagram
6a	Rearranging summing points	$Z = W \pm X \pm Y$	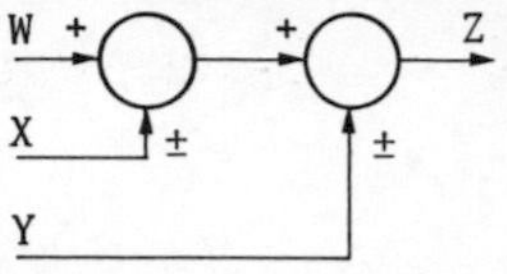 W + + Z X ± ± Y	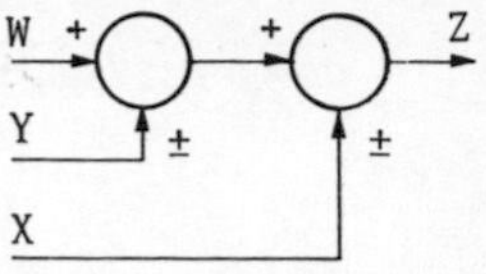 W + + Z Y ± ± X
6b	Rearranging summing points	$Z = W \pm X \pm Y$	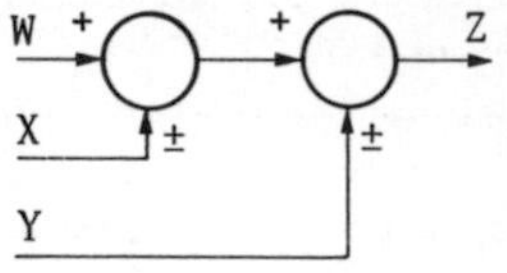 W + + Z X ± ± Y	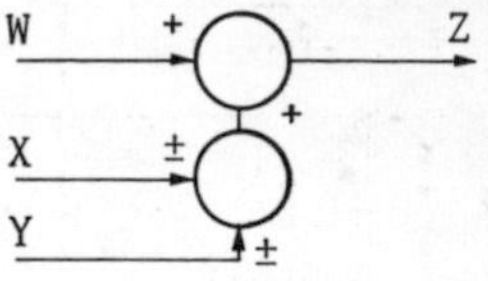 W + Z + X ± Y ±
7	Moving a summing point ahead of a block	$Z = PX \pm Y$	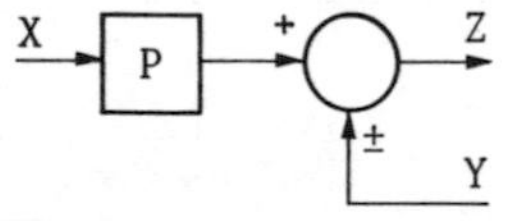 X P + Z ± Y	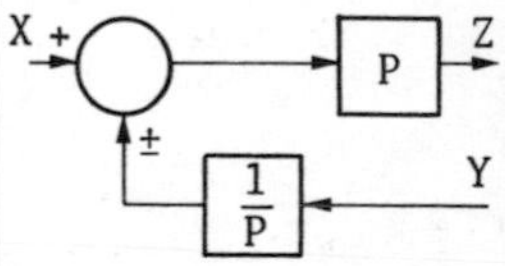 X + P Z ± $\frac{1}{P}$ Y
8	Moving a summing point beyond a block	$Z = P[X \pm Y]$	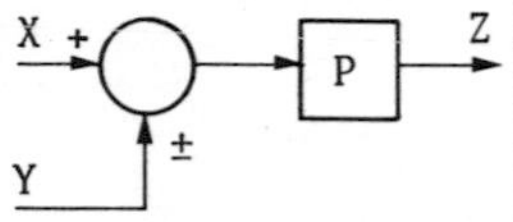 X + P Z ± Y	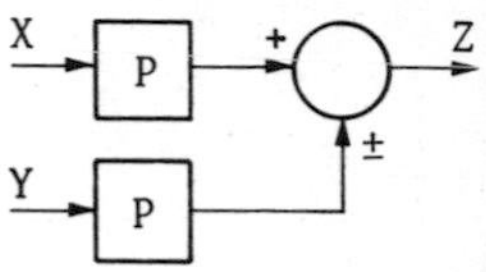 X P + Z ± Y P

9	Moving a take-off point ahead of a block	$Y = PX$	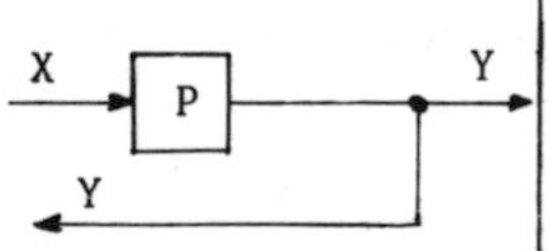 X P Y Y	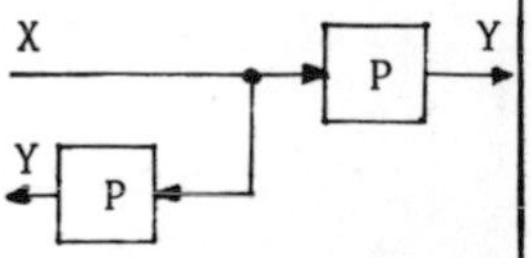 X P Y Y P
10	Moving a take-off point beyond a block	$Y = PX$	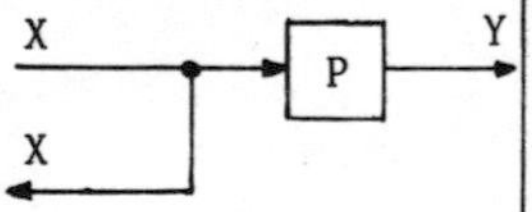 X P Y X	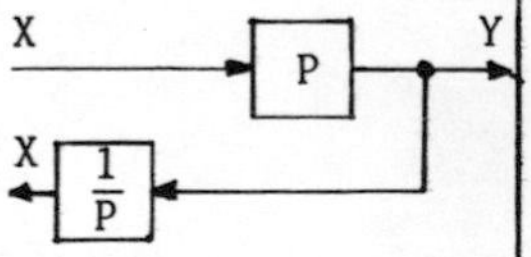 X P Y X $\frac{1}{P}$

From *Feedback and Control Systems*, by J.J. Di Stefano. Schaum's Outline Series, 1967. Used with permission of McGraw-Hill Book Company.

APPENDIX B M-circle data; N-circle data

Data for M-Circles

M	Centre	Radius	$\sin^{-1}(1/M)$
0.4	+ 0.191	0.48	-
0.6	+ 0.562	0.94	-
0.8	+ 1.777	2.22	-
1.0	-	-	90°
1.2	-3.27	2.73	56.5°
1.4	-2.04	1.46	45.6°
1.6	-1.64	1.03	38.7°
1.8	-1.47	0.84	33.7°
2.0	-1.33	0.67	30.0°
2.25	-1.24	0.55	26.4°
2.5	-1.19	0.48	23.6°
2.75	-1.15	0.42	21.3°
3	-1.12	0.37	19.5°
4	-1.07	0.27	14.5°

Data for N-Circles ($x = -\frac{1}{2}$)

α (degrees)	N	Radius	Centre y
-10	-0.176	2.88	-2.84
-30	-0.577	1.0	-0.866
-50	-1.19	0.656	-0.42
-60	-1.73	0.577	-0.289
-80	-5.77	0.506	-0.087

APPENDIX C Root locus plots

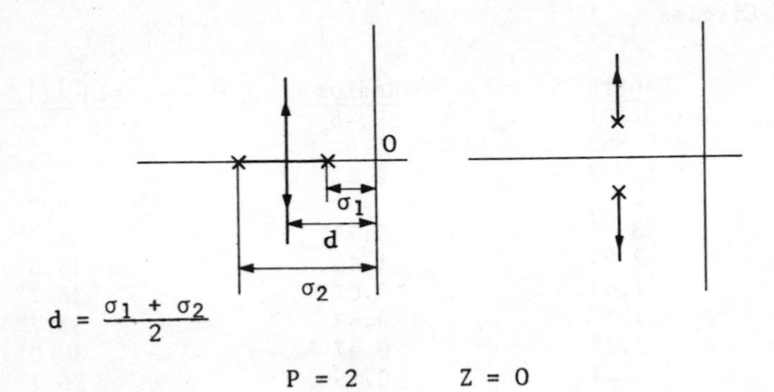

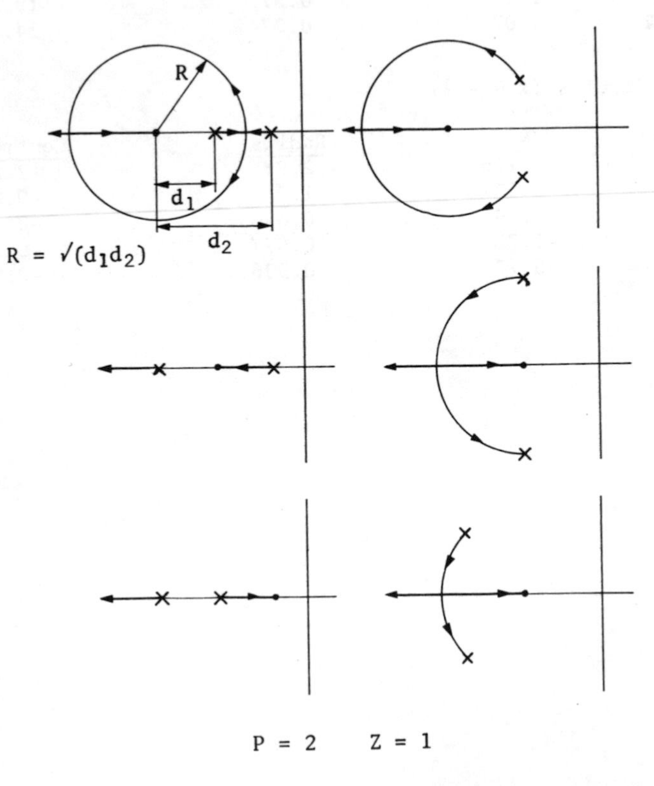

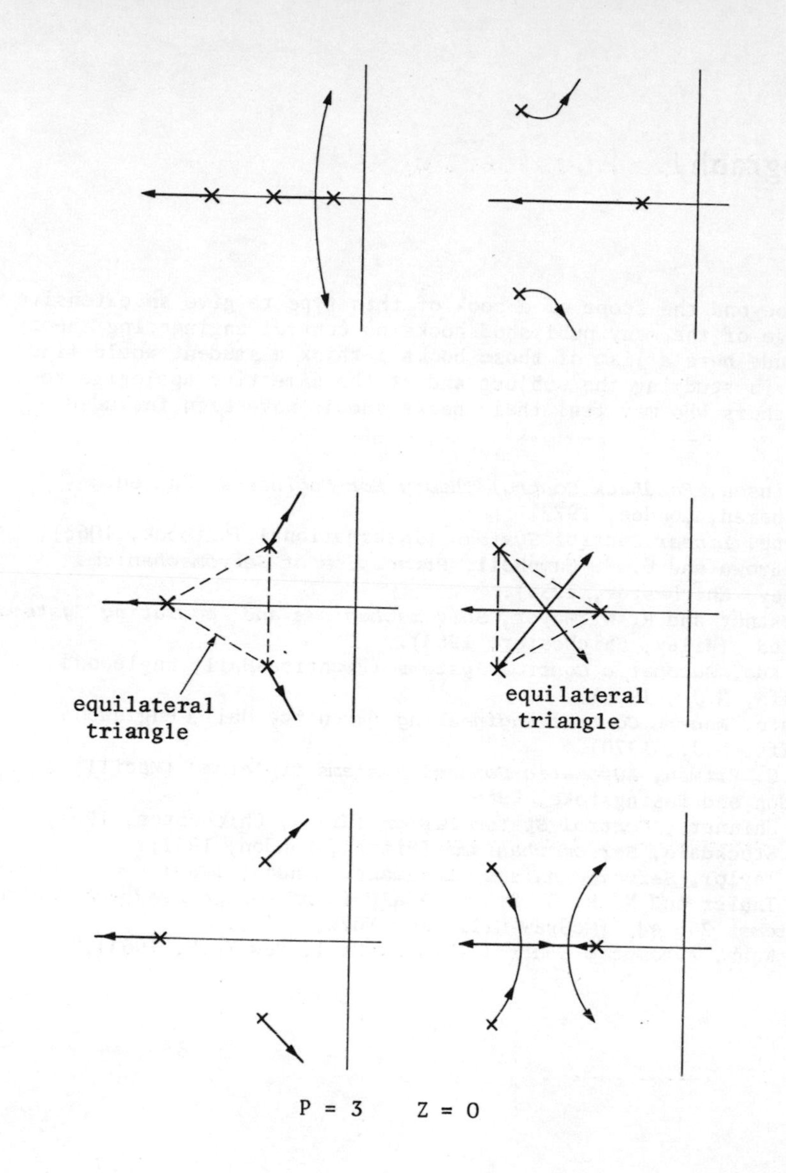

P = 3 Z = 0

Bibliography

It is beyond the scope of a book of this type to give an extensive coverage of the many published books on control engineering theory. I include here a list of those books I think a student would find useful in studying the subject and at the same time apologise to any authors who may feel their books should have been included.

P. Atkinson, *Feedback Control Theory for Engineers*, 2nd ed. (Heineman, London, 1972).

C. Barbe, *Linear Control Systems* (International Textbook, 1963).

G. S. Brown and D. P. Campbell, *Principles of Servomechanisms* (Wiley, Chichester, 1955).

H. Chestnut and R. W. Mayer, *Servomechanisms and Regulating Systems*, 2nd ed. (Wiley, Chichester, 1963).

B. C. Kuo, *Automatic Control Systems* (Prentice-Hall, Englewood Cliffs, N.J., 1962).

K. Ogato, *Modern Control Engineering* (Prentice-Hall, Englewood Cliffs, N.J., 1970).

R. J. G. Pitman, *Automatic Control Systems Explained* (Macillan, London and Basingstoke, 1966).

S. M. Shinners, *Control System Design* (Wiley, Chichester, 1964).

L. A. Stockdale, *Servomechanisms* (Pitman, London, 1962).

P. L. Taylor, *Servomechanisms* (Longmans, London, 1960).

G. J. Thaler and N. R. G. Brown, *Analysis of Design Feedback Control Systems*, 2nd ed. (McGraw-Hill, New York, 1960).

C. R. Webb, *Automatic Control* (McGraw-Hill, New York, 1964).